Successful Experiments in Indian Agriculture

Successful Experiments in Indian Agriculture

Rama Nand Sagar

CYBER TECH PUBLICATIONS
4264/3, Ansari Road, Daryaganj, New Delhi-110002 (India)
Ph.: 011-23244078. Fax: 011-23280028
E-Mail: cyberpublicationsdelhi@yahoo.com

Successful Experiments in Indian Agriculture

Rama Nand Sagar

First Edition 2009

Published by :

G.S. Rawat for **Cyber Tech Publications**
4264/3, Ansari Road, Daryaganj, New Delhi-110002 (India)
Ph.: 011-23244078. Fax: 011-23280028
E-Mail: cyberpublicationsdelhi@yahoo.com

Printed at : Asian Offset Printers

Preface

Agricultural development over the years 1949/50 to 1978/79 helped in augmenting per capita calorie and protein production by 34 and 24 per cent respectively. Wheat and rice output was the major contributor for this increase in calorie and protein production. Per capita calorie and protein production from wheat and rice touched the highest level of 1,320 kcal and 32.62 g of protein per day in 1978/79. The contribution of coarse grains in augmenting calorie and protein production was very marginal in spite of the introduction of improved strains. Per capita calorie and protein production through pulses declined by 26 per cent over the years, with the result that the ratio of cereal proteins to pulse proteins changed from 58:42 in 1950/51 to 88:22 in 1978/79. Because of the heavy tilt in cereal production, the relative prices of pulses have gone up tremendously. The impact of price rise on the consumption of pulses by the poor, which is already low. is obvious. The impact of new technology introduced for finer grains had a negative impact on coarse-grain production. This unhealthy trend is likely to affect the poor in semi-arid areas where these crops are largely grown. Seventy-five per cent of the cultivated area is rain-fed and 42 per cent of the crop output comes from these areas.

The most important impact was correcting the marginal negative trend in the per capita protein production noted prior to the introduction of the high yielding variety programme. One other important impact of the improved technology was the stability of production relating to rice and wheat.

This book is about the agricultural trends of India in past and present, the changes implemented in agriculture since far.

—*Author*

CONTENTS

1

Interfaces: Present and Future

The problems involved in bridging the wide gap between the national nutrition needs of the developing countries and available food supplies can be approached by the following three lines of action, which can be taken up simultaneously:

(a) increasing food production through better agricultural technology;

(b) ensuring effective conservation and utilization of foods through the application of modern technology; and

(c) emphasizing education in these fields, particularly the nutritional aspect.

In India, increase in crop production during the last 30 years has by and large taken place through an expansion in total as well as irrigated areas under the major food crops. Although an appreciable improvement in per hectare productivity did occur in wheat and bajra, the increase in production of other crops is largely through an expansion of the area under cultivation, or horizontal expansion. Knowing the adverse consequences of this kind of horizontal expansion on agricultural growth, the emphasis has now been shifted to vertical growth in productivity by improving all aspects of agricultural management. It is essential that the fast-growing knowledge on crop and land-use planning based on principles of economics and ecology should be transmitted to the vast number of needy and illiterate farmers.

So far agricultural scientists have used farm demonstration and other pilot projects to carry this message of science to the farmers. Other methods, such as the establishment of farm science training centres and operational research projects on a whole village, are now being implemented. While the former approach imparts the latest scientific skills to illiterate farmers as a consequence of learning by doing, the latter serves to demonstrate the value of scientific management in increasing the productivity of crops and farm animals. Such operational projects also provide an opportunity for social scientists and economists to work in close collaboration with agricultural scientists.

It is heartening to see that the recent agricultural research programme under the coordinated scheme for wheat, rice, maize, sorghum, millet, oilseeds, and pulses, is receiving greater attention than it did in the past in breeding a nutritionally superior quality of produce. Promising germplasm or genetic material with desirable characteristics is being made available by international programmes to Indian national programmes for upgrading nutritionally the quality of hybrids. In these programmes the National Institute of Nutrition has played an active part.

The recent development of high-yielding varieties of cereals and of modern agronomy to suit their cultivation has provided the basis for the new strategy in agriculture. Studies on new varieties of rice, wheat, sorghum, and pearl millet that respond to fertilizer input also lay emphasis on combining high yield with desirable nutritional quality.

In addition to fostering a fruitful co-operation between plant geneticists and nutritionists that is leading to nutritionally superior varieties of cereals and millets, this interaction has led to impressive attempts towards elimination of undesirable factors in plant foods. Gossypol-free cottonseed; erucic acid-free rape seed, and low-toxin lines of Lathyrus sativus are a few instances. It has also been possible to outwit the storage fungi affecting agricultural commodities at post-harvest stages. It is recognized that the problem of aflatoxin contamination in foods in general and groundnut in particular is a health hazard in view of its hepato-toxic and carcinogenic potential. It is therefore important that measures for preventing such contamination should receive high priority. During the last few years various approaches to this end have been vigorously pursued. The methods currently under investigation are: (a) quick-drying methods after harvest and safe conditions of storage; (b) methods for the removal of toxins by physical and chemical treatment; and (c) genetic identification of varieties resistant to fungal attack or resistant to toxin elaboration. The beginning of the last approach was made at the NIN in 1967, when our investigations led to identification of varieties of groundnut that were resistant to aflatoxin elaboration. Subsequent to this, a similar approach has been pursued at several research centres, including ICRISAT, and several promising leads have been obtained.

Looking back at the achievement of the past few decades, one feels that the progress in the application of food science and agricultural development for the betterment of nutrition, though slow, has been reassuring. The single most important challenge facing us on this front is to learn why adoption of some of the improvements already made has been slow in spite of proven advantages.

In this context, a criticism that is often leveled against the current system of integration between agriculture, nutrition, and food science is that the education component has been largely neglected. It must be recognized that there is a need to impart education at various levels, and it is equally important to realize that the kind of communication media used and the content of what is to be taught to each of the target groups will be different. Unfortunately, so far sufficient emphasis has not been placed on this aspect. There is thus a great and urgent need to carry out research on education methodology and to evolve and develop appropriate material for mass communication.

The formulation and implementation of integrated education programmes require the participation of the scientists trained in various disciplines, namely, nutrition, agriculture, home science, and food technology. The services of large numbers of young and enthusiastic scientists turned out every year in these disciplines could be well utilized for this purpose. Appropriate short-term training courses need to be started at selected centres or strengthened wherever they now exist. It is also important, as stated above, that the course content of the training programmes should cater to the needs of the target population. In view of the differences in socio-economic, literacy, and intellectual status there can be no single method by which all segments of population can be effectively reached. Let us hope that the deliberations of this workshop will take note of this and suggest measures to improve the situation.

INTERACTIONS IN FOOD TECHNOLOGY, AGRICULTURE, AND NUTRITION

The last 30 years have witnessed spectacular increases in food-grain production in India, from 51 million tonnes to more than 130 million tonnes. A sizeable buffer stock has also been built up to face the likely shortages arising out of uncertain production levels.

TABLE. Food-grain Production

Food grains	1950/51	1966/67	1978/79
Cereals	42.41	65.88	119.20
Pulses	8.41	8.34	12.16
Total	50.82	74.22	131.36

Source: DES 1975.

TABLE. Daily Per Capita Food Consumption and Recommended Levels

		Recommended level (g)	
Food materials	**Consumption (g)**	**ICMR[a]**	**Task Force on**
Cereals	434	412	376
Pulses	34	60	64
Green leafy vegetables	21	125	116
Other vegetables	71	50	61
Roots and tubers	-	87	69
Fruits	10	30	40
Milk	69	100	189
Fats/oils	12	40	39
Meat and fish (and egg)	14	30	23
Eggs	-	30	15
Sugar/jaggery	19	35	38

a. Indian Council of Medical Research

Source: Gopalan et al 1971; Gopalan and Narasinga Rao 1971; NCST 1972.

Various national projections indicate that food-grain production must be doubled to meet the needs of an estimated population of more than 950 million by the turn of this century. Since food grains constitute the principal source of calories, proteins, and other nutrients, careful efforts are necessary not only to conserve them against qualitative and quantitative losses but also to upgrade their nutritional and acceptability features. Thus, a systems approach has become crucial for integrating the production, conservation, and nutrition of food grains to get maximum benefits from national efforts.

Breeding of new food-grain varieties has been directed to increasing per hectare yields and resistance against field-borne micro-organisms and insect pests. Advances in food technology and nutrition have, however, given some insight into the desirable features that need to be considered in breeding programmes. This paper attempts to highlight some salient features of the processing, storage, and nutritive value of food grains produced in India so as to understand the interphase linkages between production and post-harvest conservation.

RESEARCH AND DEVELOPMENT EFFORTS TO RAISE FOOD PRODUCTION

TABLE. Total Cropped Area and Food-grain Production

	1960/61	1965/66	1970/71	1971/72	1972/73	1973/74
Area (in million hectares)	115.6	115.1	124.32	122.63	119.30	126.2
Production (in million tonnes)	82.3	72.3	108.43	105.17	97.02	103.6

Sources: DES 1975; Tata Services Ltd 1980.

TABLE. Yield of Important Food Grains (tonnes/ hectares)

Crop	1950/51	1970/71	1975/76
Wheat	0.66	1.31	1.41
Rice	0.67	1.12	1.25
Jowar	0.35	0.47	0.59
Gram	0.48	0.66	0.71
Groundnut	0.77	0.83	0 95

Source: DES 1978 1980.

The impressive growth in food-grain production during the last 30 years has resulted from increases in the area under food-grain crops; improvement of their per hectare yield; introduction of high-yielding varieties, particularly of wheat and rice; and an expansion of area under their cultivation, together with the provision of irrigation and other inputs.

TABLE. Area under High-yielding Varieties (in million hectares)

Crop	1966/67	1977/78
Wheat	0 54 (4.2)	15.5 (73.1)
Rice	0.89 (2.5)	15.6 (39 0)
Jowar	0.19 (1.1)	3.1 (19.0)
Bajra	0.08 (0.5)	2.6 (23.6)
Maize	0.21 (041)	1.2 (21.1)
Total	1.91 (2.3)	38.0[a] (40.3)

a. Target for 1978/79, 42 million hectares. Figures in parentheses indicate the area under high-yielding varieties out of total area under the respective crop.

Source: Tata Services Ltd. 1980.

Despite these efforts, unstable production has been a conspicuous feature even in identical areas devoted to food-grain cultivation. Hence, research and developmental efforts have focused on bridging the gap (UNCSTD 1978) between the proven yield potential and national yield average of food-grain crops.

TABLE. Gap between National Average Yields (NAY) and Yield in National Demonstrations (YND) in Farmer's Fields

Crop	Ratio of NAY to YND
Wheat	2:4
Rice	2:5
Gram, arhar, and groundout	40:70

Source: UNCSTD 1978.

Maximum productivity has been sought by judicious water-management practices, appropriate cropping systems (double, triple, and multiple) under dry and irrigated conditions, improved dryland agriculture (mulching, recycling of runoff water to provide supplementary irrigation, and choice of crop compatible with season), intercropping, multilevel cropping, and mixed farming practices (UNCSTD 1978).

TECHNOLOGICAL CONSIDERATIONS IN EVOLVING STRATEGIES FOR VARIETAL DEVELOPMENT OF FOOD GRAINS

About 70 per cent of food grains produced in India are retained for farm-level consumption and the rest moves along a chain of agencies before it reaches the consumption points. Post-harvest conservation by modern procedures is therefore a crucial need to prevent the dissipation of national efforts to raise food production levels. The incidence of bunt in wheat, chalky grains in rice, and Gibberella infection in maize, and the impairment of processing qualities as a result of pre-harvest infection have engaged the attention of scientists in recent years. The expertise in food conservation built up during the last 30 years has found increasing application, but basic information to evolve varieties with desirable storage, processing, and nutritional or organoleptic qualities is important in meeting future needs. Variable production levels in different years emphasize the need for varieties that give maximum yields during processing and suffer minimum losses during post-harvest handling and storage.

Processing

Food grains harvested from standing crops are dried and processed before cooking and consumption. Milling is used to obtain rice from paddy, dhals from pulses, and flours from wheat and millets. The wastages inherent in some traditional practices have been minimized through improved processing technology and equipment design. The physico-chemical features of food grains need attention in varietal development programmes.

Rice

About 50 per cent of all cereal production consists of rice, represented by numerous varieties exhibiting diverse per hectare yields and physico-chemical features. Research and developmental efforts have focused on the best utilization of paddy varieties grown in the country through:

a. minimizing qualitative and quantitative losses during harvest and post-harvest handling and drying stages;
b. improving the milling yields of rice by appropriate drying and milling procedures, and the development of milling equipment;
c. adapting processing procedures to improve nutrition and acceptability;
d. utilizing by-products; and
e. product development for diverse needs.

The investigations at the Central Food Technological Research Institute (CFTRI) have conclusively shown that inherent grain structure and harvest/post-harvest drying practices directly influence the milling behaviour of paddy.

Paddy is currently harvested at 16 to 18 per cent moisture level because of difficulties in getting labour, drying space, and other amenities during harvest periods. Harvesting at this stage results in shattering and sun-checking of grains ultimately reflected in heavy breakages and reduced milling yields. Harvesting at higher grain moisture levels of say 20 to 24 per cent (indicated by the presence of 1 per cent milky grains), and a controlled drying procedure with an intermediate conditioning stage, is suitable for farm level use, and also reduces milling breakages.

Breeding of crack-resistant varieties, the second approach, overcomes frequent process alterations and equipment designs. Crack-resistant and low-shattering selections from pushpa, vani. and madhu varieties have been identified, tested and released by the CFTRI for mini-test trials through the University of Agricultural Sciences, Bangalore, Further, a chalky variant of Alur sanna (a rice variety), which seems to cook into discrete, fluffy grains even without ageing, has also been identified by the CFTRI for further trials (CFTRI 1981).

The influence of physico-chemical characteristics on the storage, packing, and processing of rice has been examined as an aid in post-harvest conservation. For example, grains with low length: breadth ratio exhibit higher bulk density (and lower porosity), indicating that a given volume of round grains has a greater weight than slender grains. Similarly, the friction coefficient of grains increases with moisture level, contributing to handling and packing problems. Brown rice and highly milled rices, characterized by low friction coefficients, pack well, while intermediate milled rice, particularly from parboiled paddy, packs badly because of high friction and low bulk density (DES 1975). Yellow discoloration of rice preharvest has also been a problem in some parts of the country.

Varieties of rice with a high amylose content cook into dry and flaky products, while those with a low content into pasty products. The gelatinization temperature (GT) of starch has proved to be an important quality criterion and low-GT rices are preferred for puffed products (Bhattacharya 1979). Data such as these are extremely useful in developing varieties capable of yielding cooked products that suit the diverse food habits of Indian populations.

Breeding of varieties with a higher bran oil content, and weak or no bran-lipase activity, is also indicated in relation to the edible oil requirements of the country. Breeding for low husk content would lead to higher yields of edible material, but the storage qualities of such varieties would need scrutiny.

TABLE. Some New Hybrids and Varieties of Millets Evolved or Recommended in Recent Years

Crop	Hybrids and new varieties
Jowar	CSH-1, CSH-4, CSH-5. CSH-6, CSH-7R, CSH-8R (all hybrids).CSV-3 (370), CSV-4 (CS 3541), CSV-2 (303), CSV-5 (168), CSV-6 (604) (kharif varieties).SPY-86, FR varieties (few) (rabi).
Maize	Deccan, Ganga Safed-2, Hi-Starch, Him-123, Ganga-5, Deccan-101 (all hybrids), Amber, Jawahar, Kisan, Vikram, Sona, Vijay (all composites), Shakthi, Rattan, Protima (nutritionally superior Opaque-2 composites).
Bajra	BJ-104, BK-560, NHB-5, PHB-10 and 14, CJ-104, CK-560 (stable hybrids resistant to downy mildew), VZM-composite (resistant to mildew).

Sources. Ganga Prasada Rao 1976; Joginder Singh 1976; Murthy 1976.

Coarse Grains

Coarse grains consisting of jowar, maize, bajra, and other millets account for about 25 per cent of total cereal production. They are dryland crops capable of withstanding variable climatic conditions. Recent years have witnessed the evolution and commercial cultivation of hybrids, new varieties, and composites characterized by good yields, disease resistance, and adaptability to different cropping systems.

Coarse grains yield cooked products of harder texture than rice and wheat because of a fibrous bran layer that is relatively resistant to water permeability during the cooking process, and also of hard and horny subaleurone layers. Milling procedures and equipment (Desikachar 1976) have been evolved that remove 10 to 15 per cent of grain layers as bran to obtain grains and flours without serious impairment of nutritional qualities. Large-sized and damaged starch granules present in the grains of certain varieties of maize, jowar, and bajra are conductive to higher water uptake by flours and soft cooked products (CFTRI 1977-1981).

TABLE. Chemical Composition of Products of Milling from Jowar

Products	Crude protein (%)	Ether extraction (%)	Crude fibre (%)	Ash (%)	Calcium (mg/100 g)	Phosphorus (mg/100 g)
Unpolished jowar	11.3	4.2	1.8	1.7	49	303
Jowar (3% polish)	11.0	3.8	1.0	1.5	45	270
Jowar (7% polish)	10.7	3.2	0.8	1.3	39	276
Jowar (9.5% polish)	10.5	2.2	0 7	1.1	39	210

TABLE. Chemical Composition of Maize and Its Milled Fraction

Fraction	Oil (%)	Protein (%)	Crude fibre (%)
Maize	3.8	10.1	3.1
Debranned maize	3.9	10.4	1.6
Degermed grits	0.7	8.4	1.1
Soji (- 25 + 30BS)	0.5	10.0	0.7
Soji (- 44 + 60BS)	0.7	9.8	0.8

Source: Desikachar

In making roti (unleavened bread) from coarse grain flour, the rolling of dough into circular sheets is difficult because of the absence of gluten. The steaming of dough minimizes rolling problems, but detailed studies on the histo-chemistry of starch and its hydration features may help in identifying grains with desirable sheeting properties.

Wheat

Over 90 per cent of the wheat produced in the country is ground in chakkis to obtain flour of about 95 per cent extraction, used for making various unleavened breads such as chapati and poor). Only about 2 million tonnes pass through roller flour mills to yield maida and white flour used in bakery products. New varieties in various stages of development and cultivation are systematically screened for characteristics needed for the manufacture of bread and biscuits (Shankar and Amla 1979). Comprehensive data have become available from the Indian Agricultural Research Institute (IARI), the CFTRI, and the Food Grains Research Centre at Hapur.

A programme has been undertaken at the CFTRI to gain an insight into the protein characteristics of wheat and their relationship to the quality of chapati, the most important means of utilizing wheat in this country. Some attempts have also been made to define chapati quality by sensory methods and to correlate the data with instrumental methods of analysis (CFTRI 1977-1981). Translation of these criteria into physico-chemical characteristics could help in evolving varieties specially suited for chapati.

Pulses

Pulses are the principal means of raising the protein quality of Indian cereal-based dietaries. Pulse production has not exceeded 12.5 million tonnes, although consumption requirements for 1980181 can be estimated at 15 million tonnes (at 60 9 per capita per day for a 684 million population). To minimize the wastages in traditional milling practices and hence nutrient losses, improved procedures and milling equipment have been evolved for commercial use. This technology has been used by industry for the milling of pigeon-pea almost throughout the year. independent of the climatic conditions, which play an important role in traditional practices (Kurien 1979).

Milling conditions are also being optimized for mung bean, chick-pea, urd bean, cow-pea, kidney bean, horse gram, winged bean, and peas by suitable adaptation or modification of the technology developed for pigeon-pea at the CFTRI. Wide variations in physical properties and chemical composition of each of these pulses and their varieties are to be expected in different regions and cropping systems. These properties would influence the milling behaviour of pulses, necessitating expensive changes in technologies and equipment. Intra-and inter-varietal differences in physico-chemical properties should be exploited to upgrade the milling quality of the grain, without sacrifice of yield and duration.

Storage

Food grains in the hot and humid countries of Asia suffer qualitative and quantitative losses from insects, micro-organisms and rodents during post-harvest handling and storage. Impairment of organoleptic and nutritional qualities. and health hazards arising from insect and fungal metabolises, are well-known effects of inadequate grain protection measures and undesirable storage conditions. Food grains handled by government agencies benefit from improved storage practices, but not those remaining in farm storage.

To eliminate or minimize such losses, pesticidal chemicals and their formulations are used widely for prophylaxis or destruction of insect pests both pre- and post-harvest. Pesticide residues are monitored and regulated under Indian food laws to ensure safety to consumers. Increasing hazards from the pesticide residues on food grains accumulating in the human system. either by direct consumption or through animal-based foods, have led to an intensification of research on non-toxic insecticides, biological methods of control, and breeding of varieties resistant to pre-harvest infestation.

A mass of experimental data on the varietal resistance of different food grains to storage pests has been put out from various parts of the world. More often than not, the susceptibility or resistance of high-yielding varieties to storage pests is tested just before release for commercial cultivation. Attention to pest resistance is important at all stages of the breeding programme. Maintaining desirable genetic characteristics on a national scale of cultivation is difficult and expensive.

Sixty-seven varieties of rice have been tested for field infestation by Angoumois paddy moth (Sitotroga cerealella) and graded for susceptibility (Kittur and Patel 1972). Genetic resistance of certain rice varieties to this storage pest has also been reported from the United States (Cogburn 1977). Information on susceptibility or resistance of wheats grown in India to storage pests such as the rice weevil (Sitophilus oryzae), or khapra beetle (Trogoderma granarium), Rhizopertha dominica, and Tribolium castaneum have also been generated to help in breeding and cultivation (Bhatia and Gupta 1969; Gupta and Kalyan 1971; Chakrabarty and Mathew 1972; and Singh and Mathew 1973).

In maize, Angoumois paddy moth is found to damage varieties with a high amylose content (Fergason et al. 1970). Reports from Africa indicate that local varieties are more resistant to insect infestation than improved varieties and hybrids, because of hard kernels and complete coverage of cobs by sheaths (Adams 1977; Dobie 1977). Resistance and susceptibility of India maize varieties to Trogoderma granarium and Rhizopertha dominica have also been tested. Six genotypes (M-25-1, CSH-1, CSH-2, CSH3, CSH-4, and CSH-5) of Indian jowar have also been tested and the degrees of resistance to storage pests ascertained (Krishnamurthy et al. 1976).

Varietal resistance of pulses to storage pests has also been indicated in India. Trials on chick-pea have shown that grains of indigenous varieties G-24 and G-30 are more tolerant than new varieties to the pulse beetle (Callosobruchus chinensis) because of a wrinkled surface and tough coat (Gupta and Mishra 1970).

Continuous efforts are necessary to develop and maintain pest-resistant germplasm of different food grains so as to economize on the costs of storage and infestation control.

Breeding for Improved Nutritional Quality

Breeding for nutritional quality has mainly been directed to improving the protein content, which may or may not be reflected in increased levels of specific essential amino acids. The dwarf wheats cultivated in India contain 12.5 - 16 per cent protein and 2.21 - 2.9 9 of lysine per 100 9 protein (Deosthale et al. 1969). The protein content of improved sorghum varies from 7 to 10 per cent, and that of lysine from 0.9 to 2.6 9 per 100 9 protein. Lysine is the amino acid that limits the nutritional quality of sorghum protein; however, there is overwhelming evidence to show that excess of dietary leucine (another essential amino acid) results in a conditioned nicotinic acid deficiency. More recent studies have shown that the leucine/isoleucine ratio may be as important as the leucine level in causing pellagra. Deficiency of lysine is also relevant in bajra and ragi as in other cereals (Srikantia 1976).

Tannin present in the testa of sorghum grains at about 1 per cent level is known to reduce protein digestibility and availability by binding some of its fractions. Most Indian varieties contain less than 0.1 per cent tannin and it may not be of much significance. However, recently the National Institute of Nutrition indicated that the big-availability of iron from a variety of jowar that contained 136 mg/100 9 of tannin was lower than in one containing 20 mg/100 g of tannin. Of the common cereals, the big-availability of iron from jowar is least and iron deficiency anaemia is common among poorer sections of the population. A low tannin content in jowar is therefore a desirable breeding feature (Srikantia 1976).

Very little attention seems to have been given to the carbohydrates of food grains be they cereals, millets, or pulses, although animal experiments indicate that flatulence can occur by feeding food grains containing stachyose (CFTRI Annual Report 1980/87), and that protein quality can be affected by the level of digestibility of dietary starches. For example, feeding the field bean and some of its carbohydrate fractions to rats affected the absorption of calcium and nitrogen. Detailed carbohydrate profiles of food-grain varieties appear relevant in breeding programmes in terms of nutrition, processing and culinary practice (Srikantia 1979).

CONCLUSIONS

The impact of scientific advances in raising food-grain production levels and building buffer stocks could in future years be diluted by agro-climatic uncertainties and population growth. Accordingly, efforts have been intensified to develop sturdy varieties capable of withstanding adverse conditions, resisting infestation, and giving high per hectare yields.

Research and development in post-harvest conservation of food grains have begun to provide a fund of information on physico-chemical characteristics that influence milling yields, cooking behaviour, and sensory qualities. Inherent crack-resistant feature of paddy for better milling yields and low breakage of rice, the influence of carbohydrate make-up and starch characteristics on cooking qualities and digestibility of rice and millets, imbalances in amino acid profiles causing nutritional deficiencies like pellagra, and hard texture coupled with a wrinkled surface in grain as a deterrent against pest infestation, are some specifics that merit consideration in breeding programmes.

Commendable advances in breeding, post-harvest conservation and nutrition have been made, but only through isolated efforts. Technologies and equipment have also been developed to process varieties under commercial cultivation, but may not cope with the diversities in physico-chemical characteristics of different food-grain varieties. Constant interaction among breeders, post-harvest technologists, nutritionists, and extension specialists is necessary to integrate the developments in various disciplines and evolve a systems approach that will result in raised production levels and optimal utilization without sacrifice of nutritional or sensory qualities.

While the evolution of varieties characterized by high productivity and disease resistance will continue to have a high place in breeding programmes, genetic translation of such desirable attributes as milling yield, cooking quality, acceptability, nutritive value, and post-harvest infestation is a challenge to the scientific community.

Co-ordination between the areas of agriculture, food technology, and nutrition needs to be reflected in practice (and not just as policy) in the evolution of food-grain varieties possessing the best possible attributes of production, utilization, and nutrition. Such interaction has a very crucial role to play in the food system.

2

Generating Multidisciplinary Action

The spectre of hunger, poverty, and malnutrition continues to stare at mankind. It represents a multidisciplinary challenge of no small magnitude and therefore requires a multidisciplinary approach to find a solution. Science and technology have been able to make meaningful contributions to socioeconomic development only when they have acted in an interdisciplinary manner to solve the problems. The United Nations University has therefore recognized the value of such an approach and has given special attention to organizing activities that would involve teams of scientists (both social and natural), technologists, policy-makers and planners (including development economists) and the implementers of programmes to collectively look into the major problems of mankind and find solutions for them through co-operative efforts.

The United Nations University is doing this in the hope that the concerned disciplines will stimulate each other consciously and create a comprehensive and dynamic system capable of multidisciplinary action that could increase the pace of progress towards establishment of a more equitable and just social order in this world. This effort could convert the vicious circles in which we are caught at present into dynamic development cycles. What can agriculture, food science, and nutrition contribute towards this effort? And how can all three fields interact with each other and with other areas in order to benefit society? These questions need to be critically examined.

It is with this objective in mind that the UN University has supported the organization of interface workshops. It is hoped that the present workshop on the interfaces between agriculture. nutrition, and food science will provide an opportunity for better understanding of the whole system in greater depth and in relation to actual problems, so that more meaningful multidisciplinary solutions can be sought through co-operation between the scientific communities concerned in other fields.

The world population, which at present stands at about 4,300 million, is expected to reach a figure of well over 6,000 million by the year 2000 (FAO 1979). Agricultural production has barely kept pace even with the present need. The requirement of food, even to meet minimum need, will be nearly twice the present production level by the turn of the century and the challenges for the next century would become much greater. Food losses continue to be high and take away from mankind a substantial amount of what is produced with a great deal of inputs and human effort. The consequent qualitative deterioration of food, resulting from infestation by rodents, insects, and micro-organisms, adds to the problem of malnutrition. Prevention of these losses would increase and improve food supplies without additional demand on land, and raise nutritional standards.

The UN University looks to this workshop to provide multidisciplinary leadership, and for its recommendations that may be useful in moving forward more rapidly towards solving this global problem, which is among the greatest challenges facing mankind in the twenty-first century. scientifically, socially, and politically. In this, the efforts of the organs, organizations and bodies of the United Nations system will have an important role to play, but the real efforts needed will have to come from the countries themselves wherc the problem really exists.

THE ROLE OF SCIENCE AND TECHNOLOGY

The phenomenal rate at which science and technology continue to develop is clear from even one single indicator: 3,000 words of scientific literature are published every minute. Yet their impact on developing countries has been far from satisfactory. Only a few developing countries have benefited. These are the countries that have recognized the nature and magnitude of their problems, and have built capabilities in the form of human resources and institutions able to use the available knowledge, generate new knowledge, and further bring about its interaction with society to produce technologies that can be absorbed into the social system.

It must, however, be noted that the interaction that has taken place between science and society in advanced countries cannot always be the same as that required in the developing countries because of the fact that conditions are vastly different.

There is a need in the developing countries to bring about more meaningful and deliberate interaction between science and society if technologies are to be generated that would be really useful to them. Trained human resources that can bring about the interaction must be created, that is, people who understand not only the disciplines of science and technology, but also the interfaces involved in ensuring a multidisciplinary effort.

Only then will the results produced be such as to bring real benefit to the countries by triggering self-reliant and long-lasting processes of progress. The developing countries, which represent nearly 70 per cent of human-kind, account for barely 5 per cent of the world expenditure on research and development in science and technology.

If they have to achieve the desired results, every effort must be made to build their capabilities and optimize their impact through creation of multidisciplinary networks of co-operating institutes. The UN University is giving special attention to the need for this type of development in its Medium Term Perspective for 1982/87 (UNU 1981).

The present poverty and socio-economic stagnation has resulted in creating a condition where 65 per cent* of the world population live on about 15 per cent of the world income (Parpia 1979). The economy of these countries is primarily based on agriculture; therefore, it is through agriculture and related fields of economic activity that resources have to be generated that would contribute to overcoming poverty and bringing nutritious food within the reach of the common man. To talk of nutrition in isolation from raising income levels and increasing food supplies would be like telling the poor that if they do not have bread they should eat cake.

THE APPROACH FOR DEVELOPING COUNTRIES

When one considers the available potential, there is a great deal of hope for improving the situation. It is unfortunate that such negative aspects as the limitation of resources, population growth, and environment have received greater publicity than the positive aspects.

To exploit fully the resources available. there is an urgent need for both the political will and the provision of financial resources to build scientific capabilities. The data in table below show that there is considerable potential for better land utilization.

TABLE. Land Use, 1950 and 1960

Region	Arable as % of total		Cultivated as % of arable		Cultivated as % of total		Arable plus % of total	
	1950	1960	1950	1960	1950	1960	1950	1960
Africa	14.27	15.30	36.21	42.72	5.2	6.5	46.50	49.02
Middle East	12.87	13.91	52.11	57.88	6.7	8.1	13.06	17.34
Asia	19.03	20.78	82.06	86.17	15.6	17.9	46.35	49.60
Frontier countries (North and South America, USSR, Australia, and New Zealand)	6.88	7.75	82.85	82.96	5.7	6.4	34.27	38.59
Europe	30.79	30.98	89.02	90.06	27.4	27.9	45.63	46.10
All regions	10.73	11.73	82.74	83.99	8.9	9.9	37.35	41.07

Source: Eckholm 1976.

Different approaches are needed to increase production in developing countries from those adopted so far. At present the developing countries use barely 15 to 20 per cent of the world supply of chemical fertilizers, and much of this is used to increase production of non-food cash crops that earn foreign exchange.

Less than 20 per cent of farmers actually use fertilizers and pesticides, and most developing countries do not have a proper distribution system for these inputs. Yet it would be improper and perhaps even dangerous for them to copy the Western system. For example, in the United States nearly 300 litres of crude petroleum are used to produce one acre of corn (Pimentel et al. 1973; Steinhart and Steinhart 1974).

Even if such an expensive production system were possible, the cost of food thus raised would rise to a level where it would be out of reach of a larger number of people in developing countries. Besides, there would not be the availability of fossil fuel resources in such large quantities over a period of time. Accordingly, much greater effort is needed, on the one hand, in the area of increasing the use of biological fertilizers and developing new technologies leading to better nitrogen fixation, and, on the other, for the development of new technologies to make chemical fertilizers using solar energy and wind power. All we have done so far represents sporadic attempts to make certain nitrogen compounds (nitrites) using unconventional technologies; much more effort is required in that direction. Certain non-conventional food plants that furnish both protein and calories also need careful examination. Also, new technologies would be needed to conserve and process new foods, and to render them acceptable under different cultural conditions and food consumption patterns.

TABLE. Level of Fertilizer Consumption in Selected Countries - Kg N-P-K per Hectare of Arable Land, 1974/75

Country	Level of fertilizer consumption	Country	Level of fertilizer consumption
Afghanistan	5	Ghana	2
Bolivia	3	Hungary	216
Brazil	49	India	17
Canada	27	Iran	03
Costa Rica	130	Kenya	26
Equador	13	Netherlands	740
France	655	Republic of Korea	317
Federal Republic of Germany	311	United States of America	91

Source: FAO Fertilizer Review 1974

THE POST-HARVEST PHASES

Despite the emphasis on raising agricultural production, shortages of food continue, showing that such efforts must in future follow a different course. In addition, much greater efforts are needed on the next phase, that of post-harvest conservation and processing. Estimates of losses that occur at various points in the post-harvest system. Some of these represent intelligent estimates and indicate the need for much more data to be collected.

These losses are not only quantitative but qualitative in nature. Within four weeks of insect infestation, the protein efficiency ratio (PER) of wheat and legumes can come down substantially, while the food itself becomes unacceptable through development of undesirable metabolites.

TABLE. Estimates of Quantitative Losses during the Handling and Processing of Rice in Southern Asia

Operation	Range of losses (%)
Harvesting	1-3
Handling	2-7
Threshing	2-6
Drying	1-5
Storing	2-6
Milling	2-10
Total	10-37

Source: De Padua. 1975.

TABLE. Some Estimates of Losses in Different Areas

Region	Crop	Loss	
		Percentage[a]	Value (millions)
Nigeria[b]	Sorghum	46	
	Cow-pea	41	
United States[c]	Stored grain		US$ 500
	Packed food		US$ 150
	All crops		US$3.500
India[d]	All grains:		
	Field loss	25	
	Storage loss	15	
	Handling and processing loss	7	
	Other losses	3	
Federal Republic of Germany[e]	Harvested grain		DM 77.4
Sierra Leone[f]	Rice		41
	Maize		14
Tropical Africa[g]	All crops:		
	Storage and handling		30

a. These percentages refer to post-harvest losses unless otherwise stated. Although in most cases the figures refer to specific crops, they are sufficiently indicative to lay emphasis on the problem of food losses.

b. National Stored Products Research Institute 1952.

c. Metcalf 1962.

d. Majumder and Parpia 1965.

e. Frey 1951.

f. West African Stored Products Research Unit 1962.

g. FAO 1964.

TABLE. Mean Growth Rate and Protein Efficiency Ratio of Rats Fed on Wheat and Bengal Gram Dhal (dehusked split chick-peas) with and without Infestation (duration of experiment: four weeks; 10 female rats per group: protein level—10 per cent)

Group	Initial body weight (g)	Gain in body weight g/4 weeks	Protein intake g/4 weeks	PER[a]	PER corrected[b]
I Wheat (uninfested)	41.0	389	187	2.09	1.86
II Wheat (infested)	41 0	30.5	19.9	1.53	1.36
III Bengal gram dhal (uninfested) (chick-pea)	41.0	54.0	21.8	2.48	2.21
IV Bengal gram dhal (infested) (chick-pea)	41.0	44.8	21 5	2.08	1.85
V Wheat + Bengal gram dhal (6% + 4% protein) (uninfested)	41.0	61.3	21.5	2.86	2 55
VI Wheat + Bengal gram dhal (6% + 4% protein) (infested)	41.0	49.8	20.4	2.44	2.17
VII SMP (control)	41.0	76.9	22.8	2.27	3.00
Standard error of the mean (54 df)	+ 1.32	+ 0.07			

Results by Duncan's multiple range test at 5 per cent:

Gain in weight: All the differences are significant

Groups:	II	IV	I	VI	V	VII	
PER	1.53	2.08	2.09	2.44	2.48	2.86	3.37

Note: All means underscored by the same line are not significantly different. All means not underscored by the same line are significantly different.

a. Protein efficiency ratio.

b. PER corrected, taking the PER of SMP as 3.00.

Source: Personal correspondence with Dr. M. Swaminathan, CFTRI, Mysore

TABLE. Correlation Coefficient between Associated Factors in Sorghum

X	Y	Correlation coefficient
Moisture	Free fatty acid	0.202
	Fungi	0.170
	Total uric acid	0.339
	Apparent uric acid	0.240
	True uric acid	0.129
Fungi	Free fatty acids	0.234
	Total uric acid	0.317

	Apparent uric acid	0.904
	True uric acid	0.069
Insect count	Kernel damage	0.549
	Total uric acid	0.127
Kernel damage	Total uric acid	0.489
Toral uric acid	Apparent uric acid	0.893

Source: Majumder 1972.

TABLE. Undesirable and Harmful Metabolites Produced on Food Grains by Infesting Agents

Insect	Mould	Mites
Uric acid	Apparent uric acid	Gramine
Moisture	Moisture	Foul odour
Exuvae	Discoloration	Allergans
Chitin	Mycotoxins	
Dead insects	Thermogenesis	Pathogen vector
Frass	Musty odour	Debris
Killed-germ	Loss of viability	
Microflora		

The expected world production of cereals by 1985 will be about 1,500 million tons; even a 10 per cent loss would amount to 150 million tons.

In monetary terms, taking a modest price of US$200/ton, this would amount to US$30,000 million. Similar or even greater losses occur per ton of grain legumes, which are the main source of protein in many developing countries, and are crops of higher economic value.

This shows that an increase in research, development and technology transfer activities in the postharvest field would be more than justified even if half the losses could be prevented. Success would contribute to raising incomes and producing alternative employment in agro-industries, improvement of nutritional levels, increased use of inputs to raise agricultural production, and substantial saving on much-needed foreign exchange.

This will have a substantial impact on poverty reduction which would contribute to reduction of hunger.

It is often forgotten that very few agricultural commodities become food without some sort of processing and, in the process, they generate a great deal of employment.

Today, about US$30.000 million worth of agricultural commodities are exported by the developing countries out of which the importing countries make products worth US$240,000 million. This picture must change if trade balance is to be improved.

Another way to increase food supplies is to lay stress on greater use of primary foods. In terms of total availability of primary foods, the world produces three times its requirements. It is the consumption pattern, especially in the affluent countries, that distorts this picture, when huge amounts of basic foods are converted into secondary animal foods.

Consumers of a large amount of animal food have faced many serious health problems such as atherosclerosis and diverticular diseases. It is estimated that in countries like Canada and the United States, the annual cereal consumption per capita is 1,000 kilograms, of which barely 70 are consumed directly. The 374 million tons of cereals used by them in 1969-1971 would have sufficed to feed the entire population of India and China put together.

TABLE. Calories in Food Intake per Person per Day (1970)

	I Total calories (II + III)	II Plant calories	III Animal calories (- fish)	IV Feed calories	V Total primary calories (II + IV)
Unites States	3,300	1,869	1,431	10,017	11,886
India	1,990	1,871	109	763	2,634
Difference	1,310				9,252
Italy	2,995	2,206	789	5,523	7,729
Mexico	2,614	2,321	293	2,051	4,372
Difference	381				3,357

Source: Borgstrom 1974.

Many other types of post-harvest losses occur during the milling of rice, wheat, and grain legumes, and in the storage, handling, and processing of perishable foods such as fruits, vegetables, meat, and fish.

Inadequate utilization of oilseed proteins for human consumption also represents a major loss. Hopefully these problems will be discussed in this Workshop.

CONCLUSION

The solution to food and nutrition problems requires a sound understanding of the interface aspects, in which agricultural scientists, food technologists, nutritionists, and others concerned would constantly interact with each other to ensure a multidisciplinary attack, and work as an interdisciplinary team in a concerted manner. Only through such programmes of action can the total agro-economic system contribute to bringing about the socio-economic transformation of the developing countries, and provide the stimulus that can overcome poverty through acceleration of the development process.

3

The Improvement of Nutritional Quality by Genetic Means

Factors associated with nutritional inhibitors, such as polyphenols in sorghum, or nutritional assets, such as carotene in maize, and which can be seen by inspection, i.e. are evident, can be dealt with effectively by plant breeding.

Inhibitors and toxic factors that can only be identified by analyses or tests, i.e. cryptic factors, require laborious procedures, but levels can often be much improved. The breeding approach should always be tried. Examples are lathyrin in grass-pea or lupinin in lupine.

Improvements in cereal protein quantity and quality also require laborious procedures. In cereals such as rice and wheat, in which increased protein levels are also associated with improved lysine levels, worthwhile progress can be made by plant breeding.

Cereals such as maize and sorghum in which increased protein levels are associated with a higher prolmine content of low quality have not been responsive to plant breeding and do not warrant expensive programmes. A low-level recurrent-selection-in-populations approach ought to be tried.

Plant breeding supplies the methodology for improving crop plants. It has been likened to the process of building, although in fact it is more like replacing the coloured marble pieces in an existing mosaic by pieces of another colour, ultimately changing the pattern substantially.

In general, plant characters can only be changed one at a time, so an order of priority is required. The second requirement is that the character under improvement should be readily identifiable by the breeder. Crop improvement is a "numbers game," in which the plant with the desired combination of characteristics may be quite rare. The breeder may have to sort through a large number of plants in order to identify the very few possessing the character he needs. If such a character is evident, a careful look at each plant will suffice to choose those desired. However. few quality characters can be seen, and they must therefore be identified by other means.

Sometimes one is lucky and finds a close linkage or high correlation with an evident character: for example, kernel weight or kernel volume may be highly correlated with oil content. The breeder picks out the plants with larger grains, which greatly reduces the numbers he must handle. The oil content of these is then found by analysis, and the plants having the highest values are selected, and grown out in the next season in head-rows, preferably replicated. A sample from the whole row is taken after harvest, and analysed. There is more grain to work with, and a better analysis can usually be made of the average oil content in the row. Growing the head-row in two replications, both of which would be analysed, helps to indicate the influence of soil conditions on oil content, and permits a better judgement on whether the high oil values selected are probably due to genetic rather than environmental effects.

Often with nutritional characters, differences are not visible and there are no good correlations with evident plant characters. Then the whole screening process requires an analytical technique, with a methodology that can handle small quantities of grain (grain is available only from single plants, and most is required for seed purposes) in large numbers. Generally, nutritional factors are difficult for the breeder to work with, both because an analytical procedure is needed, and because rapid analytical techniques for small quantities of grain are seldom available. Modern technology is steadily removing the latter difficulty.

PRIORITIES

The breeder's two priorities are yield and evident quality. Both are complex. Yield involves a series of positive genetically controlled factors that enhance grain production, which need to be accompanied by a series of resistances to the yield reducers, such as pests, diseases and drought, as well as a general tolerance of difficult soil and weather conditions. Evident quality includes grain type, hardness. colour, size, texture, and flavour. All are likely to be controlled by several to many small genes each, and so require selection in large populations. Even evident quality characters are themselves quite difficult to handle. Consumer panels for tasting and cooking evaluations are needed. Further, any breeding for nutritional improvement must not result in an adverse effect on the yield, or on the evident quality, unless it brings with it an obvious advantage, a trade-off, to the farmer. He is not going to accept lower yields unless a premium price is paid for the nutritionally superior type: nor will he want to change the evident food quality if he uses it in his own home. He would only grow it for selling.

Clearly, breeding for nutritional quality is fraught with difficulties. The plant breeder must work with a (chemical) analytical technology: the new cultivar developed must be at least as good in yield and at least as acceptable in grain appearance and good preparation as existing commercial types, or else the improved food value must be sufficient to justify premium payments: the "improved" cultivar has to be economically slightly more rewarding to the grower than are existing types. This approach to nutritional improvement has therefore not been widely used for nutritional defects, which can be easily corrected by additions to the diet.

Generally, it has been important in three areas:

(a) evident characters associated with nutritional improvement;

(b) inhibitors and toxic factors; and

(c) protein quality and quantity.

Evident Characters

In the tropical crops, colour is often important. Sometimes it may be associated with tannin-like substances that help to confer some level of disease or pest resistance. Black beans are an illustration of this. Most people prefer the white ones, once farming standards no longer require the benefits that the colour confers.

Sorghum provides a good example of evident nutritional quality. Sorghum grains that have a persistent sub-coat, the testa, are high in polyphenols. Often the grains are brown in colour, and as the intensity of this pigment increases, so does the polyphenol content. Not only are the polyphenols bitter in taste, they also have adverse nutritional effects. Jambunathan and Mertz (1973) showed in rat-feeding trials that a low polyphenol sorghum gave an average gain per rat of 34.8 9 in 28 days, whereas a high polyphenol grain showed an actual average weight loss of 0.6 9 per rat over the same period. Oswalt (1973) demonstrated that sorghums with dark-coloured testes had a lower IVDMD than those lacking the dark testa. The mechanism of this nutritional inhibition is well known to this audience, as indeed it is to the African farmers who use these sorghums, and who have devised various methods, often involving soaking or germinating in wood-ash or lime, of reducing the tannin content of the grain in food preparation.

The plant-breeding task of selecting for white grain types is simple. and is also well known to African farmers. However, the birds eat the white types, so the people have the choice (where the birds are numerous) of no sorghum grain, or a bitter grain which the birds will not eat and which needs processing before the people can eat it themselves. The processes of beer-making greatly reduce tannin contents, and African beers are often important in the diet.

I used to work in the lake region of the United Republic of Tanzania. The small weaver birds, quelea, nest in the central part of the country, and their numbers reach many millions when the young fly. All the sorghum cultivars grown on the east side of

Shinyanga Region, adjoining the Wembere steppe where the birds nest, have dark grains with a high polyphenol content. Those in the western Shinyanga Region, away from the usual flight path and feeding grounds of the quelea, have sweet, corneous, white, high-quality grains. These are parts of the same district; most of the people living there are Wasukuma, and there is free interchange between the east and the west. Somebody has to get rid of the birds before the plant breeders can usefully get rid of the polyphenols.

There is one place where plant breeders can help: some of the high-tannin sorghums are very bitter and unpalatable to birds when the grain is in the milk stage, but the tannin level is much reduced by the time the grain has ripened. Provided that birds are not too numerous at harvest time, this grain reduces food-processing problems. The cultivar "Serene," released from Serere in Uganda and quite widely grown in the brown-grain areas of East Africa, is a normal high-tannin sorghum. A line from the 1965 cross 5D x 135 at Serere has now completed its field testing. Although only equal in performance to Serena for grain yield, the grain has this character of the tannin content falling markedly, as judged by taste. during ripening of the seed. It is now being multiplied and released from Serere, and it will be interesting to see to what extent it replaces "Serene."

Quinua (Chenopodium quinoa) is an important grain crop in South America, especially in the Altiplano of Bolivia and Peru. The situation there is exactly parallel to that of sorghum. The grains often contain saponins. and a breeding programme was started to get rid of them. However, it was apparent to me, when I visited our project there eighteen months ago, that saponins are present in grains in areas where the birds are numerous, and that good-quality grains without saponins are grown in other areas. (The saponins are removed by soaking bagged grain in the rivers.)

Another example of grain colour is, of course, yellow maize, in which the carotene can be a useful source of vitamin A. Not everyone likes a yellow maize, and other sources of vitamin A can easily be found in a mixed diet. The carotene levels at present known in sorghum do not warrant a serious breeding effort to include carotene in the breeding programme.

INHIBITORS AND TOXIC FACTORS

Another class of nutritional problems are posed by toxic substances, and these are very important. The haemolytic anaemia favism from eating Vicia faba beans is mentioned in every textbook. Another important one in the subtropics and tropics is lathyrism. Lathyrus, the grass-pea, is a most useful crop plant. It grows in rough conditions better than any other pulse I know. The farmer can just broadcast it into his rice field among the stubble immediately after the rice harvest: there is no land preparation, no weeding. It comes up. grows, and gives a yield under almost any condition. It can be grazed or cut, and still regenerates to give a yield of pulse. I do not know of any legume that can do a better job for the poor farmer than Lathyrus, even half as well.

It is true that the lathyrin can be destroyed by soaking and boiling the grain: but when there is pressure on the food supply, some of it gets ground up for use in bread, and then the neurotoxin begins to have its horrible effect. It is essential to get rid of the lathyrin content by breeding, and much progress has already been made here in India in achieving this, the cultivar PUSA 24 being a low-lathyrin line (Laxman Singh 1975). In Bangladesh, a pulse project supported by the International Development Research Council of Canada has taken up work on Lathyrus. An excellent range of germplasm has already been collected, including some unusually productive types, and a research worker has been here in Hyderabad for training in the techniques of lathyrin assessment.

Again, in the Altiplano of South America, lupine are a beautiful source of oil - the fields in flower are a wonderful sight - but the presence of the alkaloid lupinin requires long periods of treatment in water to prepare it as a cooking oil, and work on breeding low-lupinin lines is in hand.

Cyanogenic glycosides in sorghum furnish an example of the improvement of nutritional quality by breeding. Dhurrin is not found in ungerminated sorghum seeds, but sprouted sorghum liberates it after two days of germination, and it may be present in substantial quantities in young shoots one week after emergence from the soil. Young side branches, young tillers, and droughted and freshly rationed plants are a dangerous source of poisoning to cattle. Much progress has been made in the United States in breeding for forage sorghums and the Sudan grasses are of very low dhurrin content. Rancher was regarded as a particularly good variety (Franzke 1945), but Piper is today the most widely used of the low-HCN cultivars (Hein 1957).

Protein Quality and Quantity

The main target for nutritional improvement by breeding has been the protein quantity and quality in cereals. This was given an impetus by the concern with protein deficiency in the developing world, before it was recognized that the basic cause of protein deficiency in most situations is a food deficiency where people are not getting enough food to eat.

One attempt to develop a cereal with a high protein content that may prove to be successful is triticale, from the cross between wheat and rye. There was much fantasy in the early days, when remarkable protein contents and qualities were quoted. Little attention was paid to the fact that low grain yields are usually associated with higher protein contents, while the amino-acid profile of grains with poorly developed endosperm is often of good quality. However, now that productive types with plump grains are coming into the field, it does look as though protein content may be some 1 or 2 per cent more than in wheat, with protein quality at least as good (Hulse and Laing 1974). Triticale outyields wheat on acid soils in the tropics and subtropics in the highlands of East Africa and in South America.

The prospects of protein quality and quantity improvement in wheat itself are promising. There is a good range of variation in both protein content and lysine content. On a dry weight basis, in a large sample. mean lysine values increased from 0.33 to 0.53 per cent as the grain protein increased from 10 to 20 per cent. There was a strong positive relationship between protein content and lysine (Mattern et al. 1975).

Rice grain protein has a good amico-acid profile, with a high glutelin fraction. Milled grain usually contains some 7 per cent protein. Semi-dwarf breeding lines have been developed at the International Rice Research Institute (IRRI) which give yields comparable to those of JR-8, but with 2 per cent more protein. The higher-protein rice is nutritionally superior because of a higher level of all essential aminoacids, including lysine, in the milled rice (Juliano and Beachell, 1975).

Among the best known quality breeding programmes is that using the Opaque-2 gene in maize. A considerable amount of work has led thus far to the conclusion that Opaque-2 maize derivatives have grains that are too soft to keep well in many areas, while yields are around 10 per cent below those of the normal type. The former problem has been partly solved by the development of types with vitreous kernels, while the latter defect probably cannot be solved with the Opaque-2 gene, as the kernels of the high-lysine types are suffering from arrested development. Normal maize has around 2.5-2.8 9 lysine per 100 9 protein, while the Opaque-2 types may have around 4.0 9. It is difficult to increase total protein content usefully above 10 per cent, as the additional protein consists mainly of zein, of low nutritive value. We may conclude that farmers would do better to grow normal endosperm types, the richer farmer purchasing supplementary food rich in lysine, the poorer farmer sowing a couple of climbing been seeds at the foot of each maize plant, as he has done from time immemorial.

Sorghum is a crop greatly in need of improved protein quality. Hulse et al. (1980) state:

Recognizing the inferior nutritional quality of normal sorghum protein, the inhibition of protein by polyphenols, together with nitrogen losses that occur during domestic and small industrial processing, and probably, in storage, it is recommended that research be continued to stabilize a higher than average lysine in combination with an average (c 10% N x 5.7) protein content.

One cannot make bricks without straw: we have not yet found the straw.

In sorghum grain, the increased protein resulting from better nutrition of the plant is largely prolamine, of low nutritive value. Protein levels are low under conditions in which nitrogen is limiting at grain filling. Riley (1980) has made a thorough study of the protein situation is sorghum: the progress made in trying to use two high-lysine genes from Ethiopia, and the mutant in P-721 from Purdue. There were few grounds for encouragement. Plump seeds were unobtainable with the Ethiopian source in crosses: lysine content transferred better from P721, but yields were poor. Some figures from Riley's paper are shown in table below.

TABLE. Protein in Sorghum

Nitrogen applied	Line	Protein % (N x 6.25)	Lysine (% of protein)
20 kg N/ha	CSH- 1	4.95	3.02
	p-721	9.87	2.44
	Q.50662[a]	9.30	2.60
170 kg N/ha	CSH - 1	13.18	1.71
	p-721	15.81	2.62
	Q.50662[a]	12.49	2.48

a. Q.50662 was a "bulgy" line derived from the Ethiopian line IS 11758.

Source: Riley 1980.

The big change in lysine content with protein content of CSH-1 will be noted: the bulgy line Q.50662 had lysine levels and a response to N very similar to those of p-721. Many analyses and much effort went into this work, and there was little at the end of it showing any practical promise.

The next approach to this problem should consist of making up a population from lines with good yields drawn from diverse sources that have higher-than-average lysine contents and around 10 per cent protein when grown on land of moderate fertility. Mass selection should then be practised for ten generations, using the Udy analytical method for protein content, and progress evaluated.

GRAIN QUALITY OF SORGHUM, PEARL MILLET, PIGEON-PEA, AND CHICK-PEA

There are several components to grain quality, such as visual quality; nutritional quality, including digestibility and big-availability of nutrients; anti-nutritional factors; milling characteristics; cooking quality; consumer acceptability; and storage stability. Grain quality deserves as important a place in evaluation of new varieties as high yield and yield stability. Grain produced in a farmer's field passes through several transformations before it is consumed in the form of food. To ensure that these transformations are accomplished efficiently and with minimum nutrient loss, better interactions between different disciplines are needed, and an understanding of various aspects of grain quality becomes vital. This report describes the progress that has been made at ICRISAT in understanding some components of grain quality in sorghum, pearl millet, pigeon-pea, and chick-pea.

NUTRITIONAL QUALITY

Protein Quantity and Quality

Sorghum and pearl millet are staple foods that supply a major proportion of calories and protein to large segments of the populations living in the semi-arid tropical regions of Africa and Asia. Hulse et al. (1980) have published an exhaustive review of the chemical composition and nutritive value of sorghum and many millets. Sorghum and millet grain samples analysed in our laboratory exhibited a wide range in chemical composition . No attempt was made in these analyses to study the genotype-environment interactions.

TABLE. Seed Weight and Chemical Composition of Sorghum and Pearl Millet Grain

Constituent	Sorghum			Pearl millet		
	No. of samples	Range	Mean[a]	No. of samples	Range	Mean[a]
100-seed weight (g)	100	1.3-5.7	2.8	36	0 48-1.01	0.75
Protein (%)	34,389	4.4-21.1	11.4[b]	20,704	5.8-20 9	10.6[c]
	100	10.6-18 5	14 1			
Starch (%)	100	55.6-75 2	70.8	36	63.1-78 5	71.6
Soluble sugars (%)	100	0.8-4.2	1.3	36	1.4-2.6	2.1
Crude fibre (%)	100	1 0-3.4	1.9	36	1.1-1.8	1.3
Fat (%)	100	2.1-7.6	3 3	36	4 1-6.4	5.1
Ash (%)	100	1.6-3.3	2.1	36	1.1-2.5	1.9

a. All values except b and c are given on a dry-weight basis.

b. Based on 10,479 cultivars.

c. Based on 5,377 cultivars.

Protein content, which showed a wide variation, represents the results of analysis of samples obtained from high fertility. low fertility, unirrigated, and irrigated fields. It should be noted that the starch values reported were based on actual analysis and were not calculated values obtained by difference. The wide variability in chemical composition underscores the importance of periodically analysing representative samples of grain from the farmer's field, as normally grown, at low-fertility levels, so that a rough estimate of the availability of protein and other nutrients can be made.

A study conducted at ICRISAT using the two high-protein, high-lysine sorghums, IS-11167 and IS-11758, as parents, suggested that the high-lysine gene may not be stable in normal seed with a plump endosperm background (Riley 1980). However, further intensive study is needed to understand the reasons for this, and possibly to modify the causative factors that hamper the transfer of the high-lysine gene to a normal plump background.

Analysis of yield and protein contents in cultivars obtained from four locations indicated that, in pearl millet, it should be possible to select for increased protein content without detriment to grain yield or grain weight (ICRISAT 1980).

In the largely cereal- and legume-based diet of the Indian population, legumes serve as one of the major suppliers of protein.

From the large collection of germplasm accessions of pigeon-pea and chick-pea from various parts of the world available at ICRISAT, several thousand samples have been analysed for their protein content and a few for their proximate composition as well. The mean protein content of pigeon-pea whole seed was 20.4 per cent and that of dhal (decorticated split cotyledons) 23.2 per cent; that of chick-pea whole seed was 20.5 per cent (Jambunathan and Singh 1979, 1980). Cultivars of high-protein content could be potential sources for the development of high-protein lines.

So could the wild relatives of pigeon-pea and chick-pea: the protein content in dhal samples of wild relatives of pigeon-pea such as Atylosia, Flemingia, and Rhynchosia spp. ranged between 28.3 and 30.5 per cent (Singh et al. 1981), while some wild relatives of chick-pea showed a protein range from 25.6 to 31.7 per cent.

TABLE. Chemical Composition of Pigeon-pea, and Chick-pea Dhal[a] Samples

Constituent	Pigeon-pea[b]			Chick-pea[b]		
	No. of samples	Range	Mean	No. of samples	Range	Mean
Protein (%)	12,828[c]	13.2-26.5[c]	20.4[c]	17,679[c]	10.6 31.1[c]	20.5[c]
	8,575	15.1 - 31 5	23.2	32	20.6 30.5	25.4
Starch (%)	10	563-64.1	60.7	32	51.1-58.1	55.6
Soluble sugars (%)	10	4 7-5.7	5.2	32	4.1-6.0	5.1
Fat (%)	10	1 2-2.2	1.6	32	3.5-6.8	5.5
Crude fibre (%)	10	1.0-1.2	1 2	32	07-1.3	1.1
Ash (%)	10	3.3 4 3	3.9	32	2 1-3.7	2.9

a. Decorticated split cotyledons.

b. All values except c and d are given on a dry-weight basis.

c. Whole-seed samples.

TABLE. Protein Content and Essential Amino Acid Composition of Sorghum and Pearl Millet Grains[a]

Amino acids	Sorghum			Amino acid score	Pearl millet			Amino acid score	FAO/WHO (1973) pattern (g/16 g N)
	No. of samples	Range	Mean		No. of samples	Range	Mean		
		(g/16 g N)				(g/16 g N)			
Lysine	412	1.06-3.64	2.09	38	280	1.59-3.80	2.84	52	5.5
Threonine	29	2.12-3.94	3.21	80	29	3.17-5.66	4.07	102	4.0
Valine	29	3.84-6.93	5.40	108	29	4.38-7.67	6.01	120	5.0
Methionine + cystine	24	1.80-2.69	2.36	67	29	1.43-3.96	2.71	77	3.5
Isoleucine	29	2.85-5.05	4.17	104	29	3.70-6.34	4.56	114	4.0
Leucine	29	10.12 - 17.60	14.67	210	29	8.62-14.8	12.42	177	7.0
Phenylalanine + tyrosine	29	6.11-10.72	8.87	148	29	6.54-10.81	8.49	142	6.0
Protein (%)	412	4.60-20.25	11.98		280	6.40 24.25	12.30		

a. Determined by ion exchange chromatography.

The quality of a protein may be estimated by comparing its amino acid composition with standard reference patterns (FAO/WHO, 1973), the most limiting amino acid presumably determining the nutritive value. Generally the amino acid score is calculated as fellows: mg of amino acid in 1 g of test protein x 100.

Amino acid score = (mg of amino acid in 1 g of reference protein)/(mg of amino acid in 1 g of reference protein) x 100

The lowest score obtained for any essential amino acid may be taken as a first approximation to the probably efficiency of utilization of the test protein by children (FAO/WHO 1973).

Applying calculation to the essential amino acid composition of sorghum, it is clear that by far the most limiting amino acid is lysine (amino acid score 38), followed by methionine and cystine (67), and threonine (80). In pearl millet, lysine (52), followed by methionine and cystine (77), are the limiting amino acids.

The high level of leucine present in sorghum has been implicated in the human disease, pellagra (Hulse et al. 1980). It has been recognized that to improve the nutritional quality of sorghum and pearl millet, it would be desirable - if resources were available - to analyse the germplasm accessions of these two crops for protein and lysine contents and methods for rapid analysis are available (Jambunathan 1980).

Amino acid composition of pigeon-pea dhal samples show that the limiting amino acids are methionine and cystine (chemical score 58), tryptophan (741, followed by valine (92), threonine (95), and isoleucine (97).

In chick-pea, methionine, and cystine (67) followed by valine (89) and threonine (90) are the limiting amino acids. Surprisingly, tryptophan was not observed to be the limiting amino acid in chickpea (chemical score 115). However, since the range is large, indicating wide variability, and as no effort was made to sturdy the genotype-environment interaction in these analyses, the data should be treated with caution before drawing any firm conclusions.

Our results indicate that screening for either methionine or cystine would be sufficient, as these two amino acids are highly correlated with each other when expressed either as a percentage of protein or as a percentage of sample (Jambunathan and Singh 1982). Using rapid colorimetric procedures, we have estimated methionine content in about 500 chick-pea germplasm accessions (range 0.81-1.59; mean 1.16 9/16 9 N) and about 150 pigeon-pea germplasm accessions (range 0.70-1.48; mean 1.08 9/16 9 N) and tryptophan content in about 100 chickpea accessions (range 0.76-1.27; mean 1.05 9/16 9 N) and about 60 pigeon-pea accessions (range 0.60.95; mean 0.75 9/16 g N) (Jambunathan et al. unpublished data).

The protein quality of sorghum and millet results from the different proportions of the various protein fractions in the grain. Protein fractionation of these cereals is shown in table 5. Fractions II and III are very low in lysine concentration. In high-lysine sorghum, the proportion of these two fractions is lower and that of the first fraction higher than in normal sorghum, thus improving its overall protein quality (Jambunathan 1980).

TABLE. Nitrogen Distribution in Sorghum and Pearl Millet Grain (percentage of total nitrogen)

Fraction	Sorghum[a] Range	Mean	Pearl millet[b] Range	Mean
I Albumin and globulin	17.1-17.8	17.4	22.-26.6	25.0
II Prolamin	5.2-8.4	6.4	22.8-31.7	28.4
III Cross-linked prolamin	18.2-19.5	18.8	1.8-3.4	2.7
IV Glutelin-like	3.4-4.4	4.0	4.7-7.2	5.5
V Glutelin	33.7-38.3	35.7	16.4-19.2	18.4
VI Residue	10.4 10.7	10.6	3.3-5.1	3.9
Total	91.2-94.0	92.9	78.6 87.5	83.9

a. Based on three genotypes, M.35-1, CSH-8, and CSV-3

b. Based on four genotypes, PHB-14, BK-560, WCC75, and Ex-Bornu.

The protein quality of legumes is likewise related to the proportions of various protein classes. The proportions of fractions in the whole seed and cotyledon of chick-pea and pigeon-pea.

The methionine plus cystine contents of various cotyledon protein fractions show wide variability. Globulin is the major protein component in pigeon-pea and chick-pea. Since its content of methionine plus cystine is rather low, there is an overall deficiency of sulphur amino acids in the whole grain.

TABLE. Nitrogen Distribution and Levels of Methionine Plus Cystine Content in Pigeon-pea and Chickpea Grain (percentage of total nitrogen)

Fraction	Pigeon-pea (Hy-3C)		Chick-pea (G-130)	
	Whole seed	Cotyledon[a]	Whole seed	Cotyledon[a]
Albumin	10.2	11.4 (4.9)[b]	12.6	15.9 (5.3)
Globulin	59.9	64.5 (1.7)	56.6	62.7 (1.8)
Glutelin	17.7	18.2 (2.3)	18.1	17.5 (2.6)
Prolamin	3.0	3.5 (0.5)	2.8	2.3 (0.4)
Residue	5.3	1.8	4.9	1.0
Total	96.1	99.4	95.0	99.4

a. Embryo was removed

b. Figures in parentheses show methionine + cystine content (g/16 g N) determined by ion exchange chromatography.

Digestibility and Bio-availability

Chemical composition and amino acid composition give only an approximation of protein quality. Actual biological utilization of the protein must be determined using animals and, if possible, children. MacLean et al. (1981) conducted a feeding study on children in Peru, using two high-lysine and two conventional varieties of sorghum as the sole source of nitrogen. The mean absorption and retention of nitrogen from the sorghum diet were reported to be 46 ± 17 per cent and 14 ± 10 per cent of intake, respectively, against values of 81 ± 5 per cent and 38 ± 3 per cent, respectively, for casein under similar conditions. This indicates that sorghum is a poor source of nitrogen for children. To our knowledge, no similar studies have been carried out with pearl-millet, pigeon-pea, or chick-pea diets. Axtell et al (1981) have reported that protein from uncooked sorghum has high pepsin digestibility, which was greatly reduced after cooking; however, fermentation of flour improved the pepsin digestibility, indicating that various processing methods should be kept in view while evaluating the grain.

Mineral Elements

Nutritional studies among people of the semi-arid tropics show evidence of deficiencies in calcium, iron, and zinc, all of which elements are rendered insoluble by physic arid (Hulse et al. 1980). Sorghum shows a wide range for these mineral elements (Jambunathan 1980), but the effects of environment and other agronomic factors must be studied in detail before any conclusion is drawn. The analyses revealed interesting variations in the chemical and mineral element composition of desi (small; coloured seed coat) and kabuli (large; white seed coat) types of chick-pea cultivars grown at Hissar (Haryana, India). Desi chick-pea had a higher seed coat percentage and fibre content than kabuli. The calcium and iron contents of the kabuli seed coat, in particular, were much higher than those of the desi seed coat, and a few other minerals also showed significant differences (Jambunathan and Singh 1981).

ANTI-NUTRITIONAL FACTORS

Polyphenolic compounds (commonly referred to as tannins), physic acid, cyanogenic glucosides, etc. act as nutritional inhibitors in coarse grains. They have been reported to reduce the big-availability of protein and other nutrients, but, at the same time, polyphenols have also been shown to contribute some degree of resistance or tolerance to bird depredation, pre-harvest germination, and weathering (Hulse et al. 1980).

In legumes, in addition, inhibitors of enzymes such as trypsin, chymotrypsin, and amylase, and of flatulence-causing oligosaccharides such as stachyose, raffinose, and verbascose, play important roles. The levels of these anti-nutritional factors have been determined in pigeon-pea and chick-pea whole seed (Singh and Jambunathan 1981; Singh et al. 1982). In general, desi chickpea exhibited higher levels of these factors than kabuli. Genotypes and environments that may affect these levels were not examined in these studies. However, the low levels of these anti-nutritional factors both in pigeon-pea and chick-pea, which would be further reduced or destroyed on cooking, suggest no great need for concern on this score.

The levels of these factors and the role of polyphenolic compounds (tannins) in the big-availability of nutrients assume importance in pigeon-pea in areas where it is consumed without cooking when whole and mature, or as a developing green seed (Jambunathan and Singh 1980).

TABLE. The Levels of Anti-nutritional Factors and in-vitro Protein Digestibility in Seed Samples of Pigeon-pea and Chick-pea

Component	Pigeon-pea[a]		Chick-pea[b]	
	Range	Mean	Range	Mean
Trypsin inhibitor[c]	5.5-12.1	9.6	8.1-15.7	11.6
Chymotrypsin inhibitor[c]	2.1-3.6	3.0	6.1-8.8	7.7
Amylase inhibitor	23.6-31.0	26.4	5.0 9.7	6.8
Stachyose (%)	0.5-1.0	0.8	0.8 1.9	1.2
Raffinose (%)	0.5 0.9	0 7	0.4 0.6	0.5
Verbascose (%)	0.6-1.1	0.8		
Polyphenols (mg/g sample)	4.3-11.4	6.6	1.9-6.1	3.5
In-vitro digestibility of protein (%)	52.5-62.0	59.2	57.4-74.8	67.9

a. Based on 9 cultivars

b. Based on 15 cultivars

c. Units/mg meal

MILLING CHARACTERISTICS, COOKING QUALITY, AND CONSUMER ACCEPTABILITY

A survey carried out in Haryana, Madhya Pradesh, Maharashtra, Rajasthan, and Uttar Pradesh, which together are the major production and processing areas for pigeon-pea and chick-pea in India (Singh and Jambunathan 1980), yielded the following observations.

Pigeon-pea

Owners of 46 large-scale and 130 small-scale dhal mills were interviewed for their opinions and impressions regarding the milling characteristics of various types of pigeon-pea.

Some 130 villagers were interviewed in areas where pigeon-pea seed was reported to give higher dhal yields. Although the dhal yield was reported to vary from one mill to another, depending on the processing method used, there appeared to be a general consensus that round seeds of medium-size and greater hardness yield better dhal recovery. Village-level home processing was reported to give lower dhal recoveries than mechanically operated mills.

There were variations in recovery depending on the processing technique used. Consumers seemed to prefer local varieties grown in their own fields. There was a wide variation in colour preference because of long-time associations with a particular grain colour. Interestingly, it was observed that taste seemed to have a priority equal to or higher than cooking time.

Chick-pea

As a common practice, farmers consumed local chick-pea cultivars grown on their own farms. Cultivars with light brown testa colour were generally processed into dhal, desi types of medium seed size being preferred.

About 60 to 70 per cent were converted into fine flour, which was used in several recipes. Kabuli cultivars were preferred for consumption whole, as were bold-seeded desi types. Consumption of chick-peas in immature fresh form, or as puffed or parched products, was observed to be a common practice.

A survey of traditional food preparations from sorghum and millet was carried out in 171 villages from seven Indian states (Subramanian and Jambunathan 1980). According to the village housewives, the various foods made from sorghum and millet could be classified into the following food types: roti (coarse unleavened bread), porridge, gruel, or boiled, steamed, or fried foods.

In India, roti is the most common food prepared from sorghum and pearl millet. Studies were carried out at ICPISAT on the relationship between roti quality and physico-chemical characteristics of sorghum (Subramanian and Jambunathan 1981). The grain characteristics of 45 sorghum and 16 millet genotypes are given in table below. It was observed that sorghum flour with good dough stickiness produced an acceptable roti. The relationships between the physico-chemical characteristics and roti qualities as evaluated by a panel of tasters were tested using the methods of rank correlation and multiple regression. Results revealed that several flour components complement each other to influence roti quality. Water-soluble flour fraction, amylose, water-soluble protein and sugars appear jointly to influence and contribute to the overall roti characteristics of sorghum. In regard to pearl millet, swelling capacity of flour, water-soluble protein content, amylose, and sugars were found to be related to roti quality. Several other products that are prepared in various regions of Africa are also being evaluated at ICRISAT in co-operation with scientists located in different parts of the world.

TABLE: Physico-Chemical Characteristics and Roti Quality of Sorghum[a] and Pearl Millet[b] Grain

Factors considered	Sorghum		Pearl millet	
	Range	Mean	Range	Mean
Physical characteristics:				
100-grain weight (g)	2.3-5.5	3.8	0.5-1.0	0.75
Grain hardness (kg)/kernel[c]	3.0-11.8	6.7	1.9-3.2	2.3
Swelling capacity (final volume/initial volume):				
Grain	1.4-1.9	1.6	1.6-2.3	2.0
Flour	5.4-8.0	6.5	4.2-6.0	5.0
WSFF[d] (mg/100 g flour)	19.4-35.4	26.4	22.6-31.9	27.9
Chemical characteristics (% in whole grain):				
Protein	8.0-14.1	10 6	10.8-17.4	13.7
Water-soluble protein	0.30-0.90	0.56	1.02-1.30	1.15
Starch	62.6-74.3	58.7	63.1-70.5	66.9
Amylose	21.2-30.2	27.2	21.9-27.4	24.6
Water-soluble amylose	4.8-12.7	8.5	3.6-7.6	5.0
Soluble sugars	0.7-1.6	1.0	2.0-2.6	2.4
Reducing sugars	0.05-0.43	0.11	0.10-0.26	0.17
Fat	2.3-4.7	3.3	4.1-5.7	4.7
Ash	1.3-2.2	1.6	1.8-2.4	2.1
Tasters' panel evaluation of roti (excellent: 4; poor: 1):				
Colour and appearance	1.0-3.8	2.5	1.6-3.6	2.6
Texture	1.2-3.8	2.5	1.6-3.8	2.8
Flavour	1.7-3.4	2.6	1.8-3.2	2.6
Taste	1.0-3.3	2.6	1.6-3.0	2.5
General acceptability	1.5-3.5	2.5	1 2-3.0	2.5

a. Based on 45 cultivars

b. Based on 16 cultivars.

c. Breaking force needed (Hardness tester—Kiya Seisakusho Ltd., Tokyo Japan).

d. Water-soluble flour fraction.

Farmers and other consumers in Maharashtra stated that they prefer local sorghum to hybrid sorghum (Subramanian et al., unpublished data).

Therefore, it seems important to carry out consumer-acceptability studies on advanced breeding materials before they are considered for large-scale testing and subsequent release. In another study on six families in the Rahuri area (Maharashtra), hybrid and local sorghum flours were supplied to each family for a period of four weeks. Of the six families, four could not differentiate between hybrid and local sorghum. Further studies on larger numbers are needed to verify this observation.

Several physico-chemical characteristics of pigeon-pea and chick-pea were determined to see whether any of them could be used as a screening index for evaluating the cooking quality and cooking time of these pulses.

Important physico-chemical characteristics that showed strong relationships with cooking time included the amount of solids dispersed in cooking water, nitrogen content in the solids dispersed, water absorption of dhal, texture as determined in the Instron Food Testing Instrument (Model 1140 Highwycombe, Berkshire, United Kingdom), physic acid, and minerals such as calcium and magnesium.

Thus texture, solids dispersed, water absorption, and cooking time are some of the characteristics that should be studied to evaluate genotypes for cooking quality.

CONCLUSIONS

Considering the variability in protein content and the observed deficiency of certain essential amino acids, cereals and legumes should continue to be analysed for these constituents.

Proteins of the two cereals, sorghum and pearl millet, and the two legumes, pigeon-pea and chick-pea, were found to be deficient in methionine plus cystine.

It may therefore be worthwhile to analyse cereals for methionine plus cystine, as attempts to identify a legume high in sulphur amino acids have not so far been successful. It is important to consider the nutritional quality of the total diet, which should include the changes that occur due to processing, and so on.

Such studies may lead to the identification of a traditional product or diet that has more available nutrients for children than other products made from the same ingredients. Cooking quality and consumer acceptability criteria should be evaluated preferably in the early stages of development of cultivars in a breeding programme, and methods should be developed for rapid evaluation of these attributes.

4

The Impact of Recent Agricultural Development on Calorie and Protein Production

Agricultural development over the years 1949/50 to 1978/79 helped in augmenting per capita calorie and protein production by 34 and 24 per cent respectively. Wheat and rice output was the major contributor for this increase in calorie and protein production. Per capita calorie and protein production from wheat and rice touched the highest level of 1,320 kcal and 32.62 g of protein per day in 1978/79. The contribution of coarse grains in augmenting calorie and protein production was very marginal in spite of the introduction of improved strains. Per capita calorie and protein production through pulses declined by 26 per cent over the years, with the result that the ratio of cereal proteins to pulse proteins changed from 58:42 in 1950/51 to 88:22 in 1978/79. Because of the heavy tilt in cereal production, the relative prices of pulses have gone up tremendously. The impact of price rise on the consumption of pulses by the poor, which is already low. is obvious. The impact of new technology introduced for finer grains had a negative impact on coarse-grain production. This unhealthy trend is likely to affect the poor in semi-arid areas where these crops are largely grown. Seventy-five per cent of the cultivated area is rain-fed and 42 per cent of the crop output comes from these areas.

All the increases noticed in the output of calories and proteins through wheat and rice cannot be considered as the impact of agricultural development. To measure the impact of agricultural development. the linear trend noted between 1949/50 to 1966/67 was used to estimate the likely output M 1978/79 (if there is no high yielding variety programme) and compared this output with the actual output in 1978/79. But for reduction of 6.07 per cent in per capita lysine production, the impact of recent agricultural technology on nutrient production has been significant.

The most important impact was correcting the marginal negative trend in the per capita protein production noted prior to the introduction of the high yielding variety programme. One other important impact of the improved technology was the stability of production relating to rice and wheat.

OVERVIEW OF AGRICULTURAL DEVELOPMENT

Agricultural development over the years has helped us feed the growing millions. In the first phase of development from independence to 1966/67, there was a close race between population growth and agricultural production.

Progress in production was slow and took place largely through an increase in the cropped and irrigated area rather than through any appreciable increase in productivity. In this phase, the country was forced to import food grains during drought years. Earlier efforts to increase use of chemical fertilizers, the Grow-More-Food campaigns, and demonstrations on the cultivator's fields created an awareness, but did not motivate farmers to actually adopt the technology in a large way.

From 1961 came the era of intensive agricultural development, during which a package of services, including credit and market support, were made available to farmers, and helped them adopt the technology and augment production. Though the programme was successful wherever introduced, the larger farming community was not motivated to adopt the improved technology, presumably, because the success was not perceptible or obvious.

Later, agricultural technology from 1968 on, associated with the Green Revolution, embraces the cultivation of genetically improved strains of wheat, rice, sorghum, and maize under agronomic conditions that reveal their high yield. These high-yielding varieties and associated technology were accepted by the farming community because the yield was conspicuously higher and more economically produced than their own local varieties.

Wheat production doubled in less than five years. The infrastructure and awareness created in the earlier decades served as a launching pad for this subsequent take-off.

Over the years, extension of irrigation facilities, greater use of fertilizers, adoption of high-yielding varieties, increase in the area under improved strains, and adoption of improved management practices, have strengthened the technological base of agriculture. Sustained progress in food-grain production is still needed to improve the nutritional standards of a growing population, and the improved varieties of cereals, pulses, oilseeds, etc. have the potential to do so.

Demonstrations in the farmers' fields under the All-India Coordinated Project, the All-India Coordinated Agronomic Experiments and the National Demonstrations, have shown that yield rates could be increased, more than twofold, under irrigated conditions, the gap varying from crop to crop and state to state.

Tuber crops have also developed over the years. Critical analysis of the results of demonstrations indicate that the agricultural technology is neutral to the size of holdings (Swaminathan 1979). Production levels of these crops reflect the extent to which farmers have adopted the new technology.

TABLE. Yield Gaps in Different Crops under Irrigated Conditions

Crop	Average yield (in tonnes/ha)		Yield gap	Range of gaps in different states
	National demonstration	National average		
Paddy	5.268	2 475	2.12	1 24-3.59
Wheat	3 630	1 624	2 23	1.46-4.15
Bajra	2.623	1.112	2.35	1.72-2 80
Jawar	3.302	1 387	2 38	1.07-3.95
Maize	3.912	1 496	2.62	1 44-5.23

Source: Directorate of Economics and Statictics 1973

TABLE. Yield Gaps in Dryland Agriculture

Average yield (in tonnes/ha)

Crop	National demonstration	National average	Yield gap
Cereals			
Kharif	1,40	0 78	1 79
Rabi	1 73	0.92	1 88
Millets	0 92	0.61	1 51
Pulses	0 75	0.35	2 14
Oilseeds			
Edible oilseeds	0.60	0.34	1.76
Castor	0 52	0.23	2.26
Tapioca	4.00	1 70	2.35

Source: ICRISAT 1980

TRENDS IN AGRICULTURAL PRODUCTION

Food-grain production increased from 55 million tonnes during 1950/51 to 132 million tonnes in 1978/79. The peak output in each five-year plan shows the great impact of the new technology during the last decade.

The average annual growth rate of production during the period was 2.66 per cent. Area expansion accounts for 0.84 per cent and yield for 1.52 per cent in the growth rate. In the last decade, from 1967/68 to 1978/79, the annual growth rate was 2.77 per cent. Not all this growth is the result of new technology, area expansion accounting for 0.44 per cent and yield for 1.84 per cent.

TABLE. Trends in Peak Year Production in Food Grains in India

Five-year plan	Year of peak production	Food-grain output (in million tonnes)	Addition over the previous plan (in million tonnes)
First	1953/54	72.33	-
Second	1960/61	82.33	10.00
Third	1964/65	89 37	7.04
Fourth	1970/71	108 42	19 05
Fifth	1978/79	131.90	23.52

Source: Directorate of Economics and Statistics 1980,

Rice and wheat largely make up the increase. Out of 132 million tonnes of food grains in 1978/79. rice and wheat alone account for 89 million tonnes. The growth rates for wheat and rice were 6.02 and 2.64 per cent respectively for the period 1967/68 to 1978/79. Peak production of wheat and rice together was 52 million tonnes in 1964/65; after the spread of high-yielding varieties, this increased to 89 million tonnes, or by nearly 50 per cent. Besides growth, high-yielding varieties have lent stability to production. Even during severe drought in 1972/73, the output of 64 million tonnes was higher than the peak production of 52 million tonnes in 1964165, the highest ever obtained since independence. The same drought in 1965/66 resulted in a reduction of 10 million tonnes from the 1964/65 level (Swaminathan 1979).

Coarse grains like jowar, bajra, maize, and ragi have had differential growth rates during the last decade. During 1967/68 and 1978/79, jowar and ragi showed an annual growth rate of 2.07 and 3.98 per cent, while the growth rate for bajra was negligible, and that for maize and barley was negative. The ratio of fine grain to coarse grain was 63:37 in 1949/50, changing to 67:33 in 1964/65 and to 75:25 in 1978/79.

In regard to pulses, an important source of proteins, the introduction of improved seeds had little impact, and the output has oscillated between 9 and 12 million tonnes. During 1967/68 to 1978/79, the growth rate of production was 0.54 per cent, with an area expansion of 0.74 per cent and a negative figure for yield expansion.

Oilseed production increased from 5 million tonnes in 1950/51 to 9.9 million tonnes in 1975/76, thus registering an annual growth rate of 2.19 per cent, but in the decade ending 1978/79 the growth rate was just 1.6 per cent. Again the impact of improved seeds was not conspicuous.

Agricultural production must be viewed in the context of a growing population. Population increased from 359 million in 1950/51 to 641 million in 1978/79. An overview of the per capita production of cereals, pulses, oilseeds, etc. at the end of each plan period is given in table. Only the per capita roduction of wheat and rice has registered a 50 per cent increase, while all other commodities have either declined or, at best, kept pace. Per capita production of pulses declined conspicuously, from 70 9 per day in 1950/51 to 52 9 in 1978/79. For cassava and potatoes, per capita production increased from 30 to 76 9 during the period. The Indian diet consists largely of cereals and pulses, which account for 70 to 80 per cent of both calorie and protein intake. The proportion of pulses in total food-grain production was nearly 17 per cent in 1950/51, but declined to 9.2 per cent in 1978/79.

Uneven growth rates of individual crops has led to regional imbalances in rural prosperity, depending upon the crops grown (Singh 1980). Growth rates for various food grains and for population during 1960/61 to 1978/79 are given in table below.

TABLE. Growth Rates of Food-grain Production and Population

States	Food production[a] (1960/61 to 1978/79)	Population[b] (1961/79)
Andhra Pradesh	1.69	1.68
Assam	2.36	3.00
Bihar	1.92	1.69
Gujarat	3.56	2.34
Haryana	5.33	2.34
Karnataka	3.40	1.91
Kerala	1.39	2.19
Madhya Pradesh	1.67	2.36
Maharashtra	1.77	2.19
Orissa	1.19	2.19
Punjab	8.01	1.69
Rajasthan	2.97	2.37
Tamilnadu	1.83	1.52
Uttar Pradesh	2.79	1.68
West Bengal	2.72	2.34
All-India	2.77	2.01

a. Alagh and Sharma 1980

b. Registrar General of India 1971.

IMPACT ON CALORIE AND PROTEIN PRODUCTION

The above trends in agricultural production resulted in a considerable increase, between 1949/50 and the end of the Fifth Plan, in total calorie production, a 34 per cent increase in per capita calorie production, a large increase in total protein production, and a 24 per cent increase in per capita protein production. Before the HYVP, per capita calorie production through these major foods touched a peak of 2,304 kcal in 1964/65, but after the HYVP it was 2,446 kcal in 1978/79, an increase of 6 per cent even over the peak year.

Production of calories and proteins through food grains made rapid strides during the last decade, largely due to the success of the HYVP for wheat and rice. Per capita levels of 1,320 kcal and 32,62 g of proteins registered in 1978/79 were the highest ever reached since independence. This represents a 28 per cent increase in calories and a 42 per cent increase in proteins over the 1964/65 peak-year production (1,031 kcal and 22.90 g proteins) prior to the HYVP.

Despite introduction of improved strains, per capita calorie and protein outputs through coarse grains remained stagnant. The maximum output of calories - 5.11 kcal per capita - from coarse grain occurred in 1967/68; the drought year 1972/73 was the lowest ever recorded - 367 kcal - and this was lower than the level noticed in the drought year 1965/66, namely 396 kcal per capita.

Pulses are an important source of proteins in the Indian dietary. Calorie and protein production per capita through pulses declined by 26 per cent over the last thirty years. the decline being very conspicuous after 1967/68. Even the drought year 1965/66 saw levels of 191 kcal and 13.43 g proteins per capita, which is higher than the level of 163 kcal and 11.43 g proteins in the drought year 1972/73.

Oilseeds are an important source of calories and proteins. Per capita production of calories and proteins through oilseeds touched a level of 221 kcal and 19.05 g respectively in 1965/66. In 1978/79, this output had declined by 20 per cent for both calories and proteins.

In the case of starchy foods, per capita output of calories, though small, registered a substantial increase, from 34 to 87 kcal per capita. In regard to sugar-cane, the contribution has fluctuated at about 210 to 250 kcal per capita.

TABLE. Per Capita Availability of Calories and Proteins in Tamilnadu

Source	**Calories**			**Proteins (grams)**		
	1966/67	**1977/78**	% **increase or decrease**	**1966/67**	**1977/78**	% **increase or decrease**
Cereals						
Rice	850	1,183	(+) 39.18	15.71	21.88	(+) 39.27
Other coarse grain	384	317	(-) 17.44	10.68	8.89	(-) 16.76
Total	1,234	1,500	(+) 21.56	26.39	30.77	(+) 16.60
Non-cereals						
Starchy foods	29	70	(+) 141.38	0.17	0.34	(+) 100.00
Grams and pulses	90	97	(+) 7.78	5.53	6.05	(+) 9.40

Source: Tamilnadu Nutrition Project.

Detailed analysis of Food Balance Sheets for Tamilnadu during the period 1966/67 to 1977/78 indicate similar trends.

This analysis shows that, except for rice and wheat, the impact of recent agricultural development on calorie and protein production is not very significant. What has been gained through wheat and rice seems to have been lost through other crops. Even the increase in output, noted with regard to wheat and rice, is not necessarily due to the impact of agricultural technology. To measure the latter, the possible output during the last decade without the new technology must be surmised. This was estimated using trend lines fitted to the output data from 1949/50 to 1966/67. If the momentum gained during these 17 years had continued, the level of output of wheat and rice in 1978/79 would have been 60 million tonnes of wheat and rice can be ascribed to the impact of new technology.

On the whole, a 29 per cent increase in output has arisen from new agricultural technology in the year 1978/79. Ryan and Asokan (1977) found that 15.5 per cent of the increase of food-grain production in 1974/75 was the result of high-yielding wheat varieties in the six major wheat-growing states.

In terms of total calories and protein production through food grains, increases of 29.31 and 19.61 per cent respectively are due to the impact of new agricultural technology. In the case of lysine, the increase in total production due to agricultural technology is 20 per cent. The impact of technology on tryptophan production was highest, namely 4.67 per cent. Similarly, which gives an overview of the impact on individual nutrients, shows that there have been favourable impacts on other amino acids.

The impact on total food grains was of the order of 17.73 per cent, being highest at 28.72 per cent on wheat and rice output per capita. The negative impact on coarse grain indicates a differential bias in the promotion and adoption of new technology. In terms of per capita nutrient output, a 6.93 per cent increase in calories and 15.25 per cent increase in proteins can be ascribed to the new technology. Except for lysine, which showed a decline of 6.07 per cent, the impact of new technology on other amino acids is good, and the 23.82 per cent increase in tryptophan is salutary.

Starchy foods, sugar-cane and oilseeds also exhibit some improvement in per capita production over the plan periods. Since no conscious or intensive efforts have been made to introduce high-yielding varieties in respect of these crops, as in the case of wheat and rice, little impact is perhaps to be expected here. On the other hand, awareness of new technology could prompt use of, say, additional fertilizer when growing any new strains available. Using the same projective techniques it is found that actual production of starchy foods in 1978/79 was 17.73 million tonnes. This large increase of 69.8 per cent shows a highly significant effect of new technology on starchy foods, and the same conclusion follows from the per capita projections. However, for sugar-cane and oilseeds, the outputs, both total and per capita, are less than the projected figures based on past trends, showing that the new technology has had a negative impact on production of these crops.

NUTRITION IMPLICATIONS

Food grains, oilseeds, starchy foods, and sugar account for nearly 90 per cent of all proteins and calories in the Food

Balance Sheets. Per capita needs of calories and proteins at the physiological level for the age, sex, and occupation structure of the Indian population works out to 2,660 kcal and 55.93 g of proteins according to the most recent recommended allowances of the Indian Council of Medical Research (ICMR) (1981).

The output of calories and proteins to meet these needs would have to be 2,660 kcal, and 55.93 grams, respectively, taking into account usage for seed, feed, and wastages all along the line. Average per capita output in 1978/79 of calories through the foods considered here is 2,446 kcal, which is 92 per cent of the calories needed. The protein output per capita from these commodities is 74.20 g, much larger than the average need of the population. On the whole the calorie output is marginally sufficient while protein production is more than adequate.

The nutritional implications of the structural imbalances noticed in the production of the protein and calorie mix from different crops, and the skewing evident in distribution, require careful consideration. The protein qualities of cereals and pulses are mutually complementary in terms of amino acid composition. The food system that produced proteins through cereals and pulses in the ratio 58:42 in 1950/51 did so, in the ratio 88:22 in 1978/79. Nutritionists have harboured a fear that structural changes of this order between cereals and pulses are likely to adversely affect the quality of proteins. This fear was examined in great detail by Raya and Asokan (1977), who found, on the aggregate, that but for reduction in lysine production by 5.8 per cent, the impact of the Green Revolution on the overall calorie and protein production was significant and positive in the six major wheat-growing areas.

It was argued that in terms of the trade-off between the higher output of calories and proteins through raising high-yielding varieties, the small reduction in lysine production should not worry nutritionists. Similar analysis on an all-India basis shows that there is, in fact a 21.4 per cent increase in the aggregate total lysine production resulting from new technology. However, analysis of the trend in per capita production reveals a reduction of 11.12 per cent in lysine.

Except for this reduction, recent agricultural technology has had a significant impact on the per capita production of food grains, calories, and proteins.

The marginal declining trend in per capita production of proteins noted in the years 1949/50 to 1966/67, became corrected to a significantly positive trend during 1967/68 to 1978/79 following the HYVP, and this impact is important from the nutrition angle. The 4:1 ratio of cereal protein to pulse protein in 1978/79 is in keeping with the recent recommendations of the ICMR (1981). Because of a heavy tilt towards cereal production, the relative prices of pulses have gone up tremendously. Any price rise will obviously reduce the already very low consumption of pulses by the poor.

If lower deciles of the population have to get the recommended 4:1 ratio of cereal and pulse proteins at the retail level, the ratio at the production level will have to be different, in keeping with the relative prices of cereals and pulses. Such a drastic reduction in the ratio is also not conducive to soil health. Nitrogen fixation of these leguminous crops helps in saving non-renewable energy for production (Swaminathan 1979). To achieve a healthy overall input-output of nutrients, the mix must be optimal with reference both to soil nutrients and human nutrition.

The proportion of protein and calorie output per capita derived from coarse grains like maize, jowar, and bajra to the total of calories and proteins from food grains was 30 per cent in 1950/54, and dropped to 22 per cent in 1978/79. As has been pointed out, the new technology has had a negative impact on coarse-grain production. Actual per capita production in 1978/79 was 130.11 g against an anticipated figure of 134.52 g. This unhealthy trend is likely to affect, especially, the poor in semi-arid areas where these crops are largely grown. Of the cultivated area, 75 per cent is rain-fed and 42 per cent of the crop output comes from these areas.

Such areas have been plagued, for centuries, by periodic drought, floods, soil erosion, instability of production, drinking water scarcity, unemployment, and other forms of human suffering (Swaminathan 1973). Agriculture is a gamble over about 101 million hectares where most of the coarse grains are grown in competition with other commercial crops like groundnut and cotton. The consumption patterns of the poor reveal that the share of calories from coarse cereals, out of the calories from all cereals, was 42 per cent and the share of proteins, 50 per cent. These ratios varied from as low as 2 per cent in West Bengal to 90 per cent in Rajasthan, and 85 per cent in Gujarat and Maharashtra (Jodha 1973). As to nutrients, these grains are as good as rice and wheat. Ragi, rich in calcium, is good for hard-working peasants. The importance of these grains has been realized, and ICRISAT is developing new technologies to put them on a par with other fine grains. New technology alone is not adequate and it should have all the support of extension programmes, government pricing policy, procurement, storage, etc. which helped to introduce the new technology for rice and wheat. There is need for integrated programme planning and policy formulation in the promotion of different crops. A judicious crop mix will protect farmers from the risks attendant on vagaries of the monsoon as well as optimize the use of land and inputs. Analysis by state of the impact of new technology would aid integrated food and nutrition planning.

CONCLUSIONS

Agricultural development over the years, 1949/50 to 1978/79, helped in augmenting per capita calorie and protein production by 34 and 24 per cent. respectively. Wheat and rice output was the major contributor to these increases. Per capita production from wheat and rice touched the highest level of 1,320 and 32.62 9 of proteins in 1978/79. The contribution of coarse grains to augmenting calorie and protein production was very marginal, despite the introduction of improved strains. Per capita calorie and protein production through pulses has declined by 26 per cent over three decades, and as a result the ratio of cereal proteins to pulse proteins changed from 58:42 in 1950/51 to 88:22 in 1978/79. Though there has been an overall increase in calorie and protein output, the structural imbalances and their regional distribution need to be examined from a nutrition angle.

All the increases in calorie and protein output through wheat and rice cannot be considered as the impact of agricultural development. To measure this impact, the linear trend noted between 1949/50 to 1966/67 was used to estimate the likely output in 1978/79, if there had been no high yielding variety programme, and this output was compared with the actual output in 1978/79. Except for a reduction of 11.12 per cent in per capita lysine production, the impact of recent agricultural technology on nutrient production has been positive and significant. The most important impact was in correcting the marginal negative trend in per capita protein production that appeared to have occurred prior to the introduction of the high-yielding variety programme.

5

Cropping Systems for Upland Areas

One of the most important aspects of formulating a cropping system for an upland situation (whether impaled or rain-fed) is the selection of crops and varieties that are suited - with reference to their yield and quality - to the region. Foods crops needed to meet our requirements have been classified into three broad categories, e.g. energy food (cereals), body-building food (grain legumes) and regulatory food (vegetables and fruits).. Bearing in mind food habits, nutritional requirements and nutrient availability (which is much below the requirement level in the country), a scientific approach is needed for the augmentation of productivity in the uplands growing the three categories of food crops. In the absence of the possibility of increasing the cultivable area, the only way out is to increase productivity per unit of area and time. While a cropping system on irrigated uplands can be manipulated in a variety of ways, the crops and the system of cropping on rain-fed uplands afford less flexibility, in so far as they necessarily depend, primarily, on the effective utilization of the precipitation received before and during the cropping season. Climat and edaphic factors contribute significantly to the success or failure of the harvest.

Intercropping moth bean, guar bean, and mung bean with pearl millet; pigeon-pea. black gram, groundout and soybean with sorghum, has been suggested for normal rains, while intercropping mung bean and pigeon-pea with sunflower/maize/castor has been advised for aberrant weather situations (late monsoon) in and areas with a rainfall of less than 700 mm.

Collection of run-off water in storage tanks. water harvesting in situ, surface mulching, and establishment of optimum plant population through seeding at optimum depth of soil at appropriate sowing time are the proven technologies for successful cropping in and areas. Application of farmyard manure on seed furrows to avoid crust formation and reduce seeding mortality, along with increasing moisture-holding capacity, should provide invaluable aids in crop establishment. In such areas, even in favourable and normal years, sequence-cropping is advisable, e.g. pearl millet/sorghum (for fodder), or pearl millet and cow-pea (for fodder), followed by mustard. Alternatively, short duration varieties of black gram or mung bean, followed by chick-pea/mustard/raya/safflower. The cropping strategy for moderate rainfall areas (700-1,100 mm rain) may be intercropping systems such as maize and green gram/soybean; cotton and green gram /for the north and north-western region); groundnut + red gram, cotton + green gram (for the central region) and groundnut/finger millet + red gram (for the southern region) and a sequence-cropping system, such as maize/early rice followed by chick-pea (for north and north-western region); pearl millet/sorghum followed by chick-pea or black gram followed by wheat or green gram followed by sorghum (for central region); sorghum-cow-pea, green gram - sorghum/sunflower (for southern region).

Areas with rainfall more than 1,100 mm should adopt intercropping of pigeon-pea, groundnut, mung bean, soybean in main crops of maize/sorghum/early rice. Proper crop geometry between intercrop and main crop has to be worked out on a regional basis. For the eastern region, two or three rows of rice alternated with one row of mung bean or three to four rows of rice with one row of pigeon-pea, have been optimum. Intercropping legumes in cereals have shown increased yield of grain as well as protein and lysine. In soils retaining adequate moisture and nutrients, double cropping systems like maize-toria/wheat, rice-chick-pea and soybean-wheat, have been suggested.

It has been observed that suitable crop combination increases productivity not only in quantity but in nutritive quality as well. It has been reported that mixing whole wheat meal with green gram (mung bean) in the proportion of 90:10, increased the chemical score of protein in the combined diet to 71 per cent, as against 53 and 47 per cent for wheat and green gram alone. Cereals, like wheat and barley. in combination with other pulses, like cow-pea, lentil, red gram, pea. and chick-pea, have given similar results.

Cropping systems are dependent on their physical and socio-economic environment. The productive base of any cropping system is plant growth, which is influenced by environment and management. The environment includes physical factors related to climate (rainfall, temperature, and solar radiation), land (soil types, location, and topography) and socio-economic resources specific to costs, like the availability of production inputs, product prices, marketing facilities, and preferences for certain food and management practices. Overall, the cropping pattern performance is the continuous function of available resources.

Cropping system design employs certain criteria, like the yield of the component crops within the pattern, which should be economically viable and biologically stable. The economic resource requirement of the system should be such that the input of labour, power, seed, fertilizer, and plant protection can be met in time, adequately, and economically. Physical variables, such as rainfall, temperature and soil types, influence the system classification and define the ecological zones. An ecological zone is one across which a relatively uniform potential of crop growth is a common phenomenon. It would not amount to an oversimplification to consider the cropping system design approach as one that involves the matching and linking of the environmental conditions for crop growth with the management requirement of the crops in combinations that bring about their best potential. To design each cropping system, it would be necessary to identify environmental, physical, socio-economic, and biological variables that relate to crop growth and management. However, the determination of crop requirement and the classification of environmental parameters may not always be very precise, and the design may be developed on the basis of probabilities. Cropping system design is not a mechanical or automatic process; it must be developed from an understanding that agricultural practices, that are to be evolved and adopted successfully across an ecological zone, are an evolutionary process, in which a minor adjustment of the system should provide the prospect of exploiting any potential to the optimum.

Out of the 140 million hectares of total cultivated area in the country, about 25 per cent is irrigated. Except for the area under lowland rice, the remaining cultivated land may be categorized as under upland management - little is medium land. Such crops are grown on the soil moisture received from rainfall, the system is known as rain-fed farming. For successful adoption of crops and cropping system, the primary consideration is that the conserved soil moisture is a limiting factor.

Since there is little more land to be brought under cultivation, the increasing demand for food in the country must be met through increased productivity of the existing land per unit area and per unit time. The productivity of rain-red uplands being low, judicious crop planning with scientific management aimed at exploiting resources to the maximum, would be the primary approach. The production objective would be not only the quantity but also the quality of the produce specific to the nutritional needs of the people.

CROPS AND NUTRITION

Crops meeting the nutritional needs can be categorized. Energy-rich foods include cereals (rice, wheat, barley, millets), pulses, sugar, and oilseeds. Body-building foods are crops like grain legumes, rich in proteins.

Regulating foods are fruit and vegetable crops. Other sources of nutrients, such as cattle, poultry, and aquaculture. though significant, are beyond the scope of this paper.

Data recorded for 1978 appear to show that the per capita consumption of pulses was only 45 g per day, while the total area covered by pulses was 23.5 million hectares providing 12.1 million tonnes of legume grain. WHO/FAO has recommended consumption of 104 g pulses per capita per day. Based on a more modest estimate of 60 g, our pulse needs would come to about 24 million tonnes. Thus, the productivity of pulses has to be doubled, and this is attainable by following a cropping system on the uplands with cereals and legumes in sequence or intercropped. Successful practice of intercropping has opened new vistas for rain-fed uplands where soil moisture is the usual constraint.

A combination of cereals and pulses in the Indian diet has a complementary relationship in meeting the essential amino acid needs both of lysine and the sulphur-containing amino acids, like methionine. Mixing whole wheat meal with green gram (mung bean) in the proportion of 90:10 increased the chemical score of protein in the combined diet to 71 per cent, against 53 and 47 per cent for wheat and green gram alone. Cereals like wheat and barley in combination with pulses, like cow-pea, lentil, red gram, pea and chick-pea, give similar results. Combinations of cereals and pulses give a chemical score for proteins which compares favourably with those of animal protein (Jain 1980).

The choice of crop and the cropping system for rain-fed uplands depend, in the main, on effective utilization of the precipitation received during and before crop cultivation. Management of these lands has been developed with the prospect of utilizing the entire rainfall received on the land. The areas receiving varying amounts of precipitation have been classified into three categories: areas of low rainfall (below 700 mm), medium rainfall (700-1,100 mm), and high rainfall (above 1,100 mm). Both the amount of precipitation and management decide the mono-, inter- or double-crop systems to be followed. Besides these, other relevant climatic factors are temperature, sunshine, and humidity, besides soil types and soil condition. Thus, the whole system is location-specific. Socio-economic considerations, like cost, availability of production inputs, marketability and consumers' preferences for the produce, influence the crop-growing decisions within the cropping system.

CHARACTERIZATION OF UPLANDS

Land topography and depth and type of soil vary greatly in the uplands. A higher elevation, uneven topography and, occasionally, light-textured soil condition, will limit soil-moisture storage and hence its availability to the crop. Some uplands having sandy loam or lateritic soils have poor native fertility. They are characterized by low organic matter and nitrogen status, poor availability of phosphorus and iron, and poor retention capacity for both moisture and nutrients. Rain-fed uplands are located in zones of both low and high rainfall. They may be subject to periodical drought, due to untimely and irregular distribution of rain, during the crop season. Socio-economic factors vary. As a result, upland cropping, particularly rain-fed farming, shows wide diversification within the country.

The rest of this paper will discuss the suitability of crops (cereals, legumes, or oilseeds) and, or, cropping systems as influenced by the rainfall pattern (low, medium, or high) and interaction with input management (low or high), which is a consequence of socio-economic resources. The illustrations may not be exhaustive but are fairly representative.

Low Rainfall Zones (average rainfall below 700 mm)

Crop production strategy in low-rainfall areas includes effective utilization of rainfall by water-harvesting techniques, such as proper use of simple terraces, contour furrows, and bunding, which permit higher water retention by arresting the loss of run-off water. Jowar, bajra, or maize during normal rainfall (monsoon) by kharif intercropping with oilseeds or legumes, or through prospective environment sequence-cropping during rabi on conserved soil moisture, show good prospects with adequate and timely input management. Even so, alternative cropping strategies must be developed for aberrant weather conditions. particularly, in areas characterized by light soils coupled with scanty precipitation and an arid climate. The choice of crops or crop varieties that can mature with the available moisture received through precipitation would be an important part of the approach. In areas of low rainfall combined with a low management input, one prevalent technology is mixed cropping or intercropping. This practice is fairly helpful in normal as well as aberrant situations. In years of normal rainfall, the system will increase total productivity from a unit land area, and in abnormal years it would act as an insurance against complete crop failure.

Table 1 shows that double rows of moth bean, guar bean, and mung bean planted in the interspace of paired rows of bajra (30/70 cm) give an additional yield of grain legumes without affecting the yield of the principal bajra crop (Josh) and Singh 1981). Not only was total grain production higher, but the monetary benefit was even more favourable, because legume grain prices are appreciably higher than those of the cereal.

TABLE. Yield of Base and Intercrops in Cropping System

Treatment	Yield in kg/ha		
	Base crop	Intercrop	Total grain
Uniform row of bajra	1,447	-	1,447
Paired row of bajra	1,385	-	1,385
Paired row of bajra + 2 rows of moth bean	1,274	381	1,655
Paired row of bajra + 2 rows of guar bean	1,249	381	1,650
Paired row of bajra + 2 rows of mung bean	1,157	458	1,615

TABLE. Grain Yield from Intercropping Legumes with Sorghum as Base Crop

tment	Grain yield (kg/ha)					
	1973/74			1974/75		
	Base crop	Inter-crop	Total	Base crop	Inter-crop	Total
ıum	717	-	717	2,814	-	2,814
ıum + mung bean	1,148	216	1,364	2,750	265	3,015
ıum + pigeon-pea	1,405	389	1,794	2,631	902	3,853
ıum + groundnut	1,362	170	1,432	3,130	426	3,556
ıum + soybean	-	-	-	2,690	299	2,989

Source: Srivastava and Singh 1980.

Similar, is the observation with sorghum as a base crop, and legumes (pigeon-pea, black gram, groundnut, and soybean) as intercrop . This cropping system yielded an increased amount of total food, besides a higher nutritive value. Among the different intercrops, pigeon-pea (var. Prabhat) gave the best performance (Srivastava and Singh 1980).

Under less favourable soil moisture situations, like delayed onset of the monsoon, the strategy lies in selecting crops constituting the intercrop combination with the ability to sustain soil moisture stress. Intercropping, using sunflower + mung bean, sorghum + pigeon-pea, maize + castor or mung bean or black gram in specific situations, has shown promise. Any further success depends on using management techniques to utilize soil, water, and fertility resources so as to provide a favourable environment for plant growth.

Well-known soil and water conservation methods, including the collection of run-off water in storage tanks, and using the water as a life-saving measure when the crop faces drought, needs no emphasis. Both water harvesting in situ using the inter-row method. and surface mulches have proved useful and economical.

Non-monetary inputs, like optimum time of planting and the depth of seed placement when a dry-seeding technique is followed, need to be adopted precisely. The depth of seed placement must be so regulated that the pre-monsoon rain, which activates the seeds to germinate, has to permeate to such a soil depth that the total moisture, held in the soil profile following rainfall, is enough to support seedling growth and survival, till the regular monsoon sets in.

Further, an optimum plant population must be maintained by adopting a suitable seed rate and row spacing. Application of farmyard manure in the seed furrows has been found useful in avoiding soil crust formation, which causes seedling mortality.

TABLE. Grain Yield of bajra Grown under Different Types of Management

	Yield in kg/ha	
Treatment	**1978**	**1979**
Bajra sown by pore method	706	389
Bajra sown with ridge and furrow method	1,316	834
Mixed seeding of bajra and legume	1,193	634
(legume removed after plant establishment)		
Gap filling of bajra using 25-day-old seedling (during rains)	1,655	901
Bajra sown with farmyard manure applied on seeded row	1,767	1,019

Source: Joshi and Singh 1981

Maintenance of a weed-free environment is an essential and integral part of management to sustain the productivity of crops in an arid region. Tillage and harrowing keeps weeks under check, and makes a thin layer of soil mulch which is helpful in moisture conservation.

In cases of early withdrawal of monsoon, the following practices may prove useful: Thinning of stand by removing alternate rows of bajra to be used for fodder. Removal of lower leaves in maize and sorghum to reduce transpiration losses.

In an intercropping system, removing moisture - sensitive crops for use as fodder, while maintaining resistant ones to maturity to produce grain. Mid-term corrections like harvesting of maize and sorghum for fodder, and early sowing of rabi crops like mustard, safflower, chick-pea, and sunflower.

Under delayed monsoon conditions, practices that may be helpful are: planting of 20-25-day-old seedlings of bajra rather than sowing seed; replacing sesame with sunflower, and bajra with castor or guar bean; early sowing of sunflower, chick-pea. mustard, raya, and safflower on conserved soil moisture in the rabi season.

When the rainfall is normal and regular, and the area has a soil type with good soil moisture and nutrient retentive capacity, a favourable environment exists for supporting sequence cropping.

Successful cropping systems are: bajra/sorghum (for fodder)-mustard; bajra +cow-pea (for fodder)-mustard; black gram-chick-pea/mustard; mung bean-raya/safflower; and bajra-raya.

Using double cropping, forage yields of over 50 grams per hectare could be obtained by growing fodder (sorghum/bajra) and harvesting it before recession of the monsoon. After harvest of the fodder crop, seeding mustard (T. 59) on residual soil moisture gave a good harvest yield of 1,1 tonnes per hectare d/ha (Rathore 1980).

Situations occur where the monsoon sets in at the normal time, but it follows a long dry spell, with late subsequent rain again later. Here, early sowing of mustard (T. 59), safflower (N-62-8) or chick-pea (RS 10) helps to exploit the situation, yielding a good harvest of winter crops under conserved soil moisture.

Moderate Rainfall Zones (average annual rainfall 700-1,100 mm)

In the moderate rainfall zone, in the rainy (kharif) season, maize or cotton may be taken as a base crop. Intercropping with legumes like green gram, groundnut, and red gram, has been successfully followed in the north, north-western and central regions of the country. Since cotton or groundnut forms the base crop, intercropping enables a high cash earning. In the northern zone, winter (rabi) crops are also being grown on conserved soil moisture, either pure or intercropped, e.g. barley + chick-pea, barley + pea.

Some proven intercropping systems successfully adopted in different regions of the country under rainfed situations are as follows: north and north-western region: maize + green gram/ soybean; cotton + green gram; central region: groundnut + red gram; cotton + green gram; southern region: groundnut/rag) + red gram; and north India in the winter season: barley + chick-pea/peas/mustard.

The beneficial effect of intercropping legumes (groundout and cow-pea) in cereals (maize and sorghum) has been established in rain-fed areas, both in terms of grain yield and land equivalent ration (Prasad and De 1980).

TABLE. Yield and Nutritional Efficiency of Some Intercropping Systems

	Yield (kg/ha)			Estimated yield of protein (kg/ha)	
	Base crop	Inter-crop	Total	Total	Lysine
ɔ (1973/74/75)[a]					
	4,764	-	4,764	476	11.4
igeon-pea (Hy-3A)	4,737	1,226	5,963	744	28.8
ɔigeon-pea (Hy-3A)	4,299	1,239	5,538	703	27.8
-pea (3:1 ratio)	4,152	1,259	5,411	692	27.6
'op (1973/74)[a]					
	1,056	-	1,056	274	9 9
pigeon-pea (Hy-3A)	1,100	1,487	2,587	603	30.6
ɪn-pea (3:1 ratio)	1,017	1,110	2,127	508	25.1

Source: Reddy, Tarhalkar, and Rao 1977.

Table above demonstrates that intercropped sorghum + pigeon-pea gave an increased total grain yield, ranging from 15 to 25 per cent, over a pure sorghum crop. From the nutritional viewpoint also, the total protein yield increased from 45 to 55 per cent in value. Intercropping of sorghum and pigeon-pea resulted in an increase of 140 to 150 per cent in lysine yield over the pure sorghum crop. Considering groundnut as the main crop, pigeon-pea as an intercrop gave an increase of 100 to 150 per cent in total production. The groundnut + pigeon-pea system was found to be superior in terms of both total protein yield as well as quality (lysine yield), registering an increase of 100 and 200 per cent, respectively, over a pure groundnut crop (Ready, Tarhalkar, and Rao 1 977).

In a moderate rainfall zone, soils with good moisture retention capacity may sustain two crops in sequence with adequate moisture and high management inputs. Table below shows some possibilities. With an appropriate choice of crops, a two-crop sequence can be successful in rain-fed areas. Hybrid bajra (BJ104) or short-duration fodder in the rainy season (kharif), and safflower/mustard/chick-pea during the winter season (rabi) have been found successful in Jhansi (Patil and Ali 1980). Inclusion of bajra in the sequence made double cropping a feasibility in other areas too, like bajra-mustard (Hisser and Agro); baira-chick-pea (Varanasi), bajra-wheat (Jammu), bajra-cluster bean (Jodhpur).

TABLE. Suitable Sequence Cropping for Different Zones of India

Region	Rainy season *(kharif)*	Winter season *(rabi)*
North and north-western	Maize Rice (early)	Chick-pea Chick-pea
Central	Bajra/sorghum Black gram	Chick-pea Wheat Sorghum
Southern	Sorghum Green gram	Cow-pea Sorghum/sunflower

Source: Annual Reports 1970-76.

High Rainfall Zone (mean annual rainfall above 1,100 mm)

As in the moderate rainfall region, intercropping systems with maize, sorghum, and rice (early) as the base crop, and legumes like pigeon-pea, groundnut, mung bean, and soybean as the intercrop, have proved successful in higher rainfall areas.

Intercropping legumes (mung bean, groundnut, and pigeon-pea) in upland rice in the kharif season in the ratio of 3:1, i.e. three lines of rice (20 cm apart) alternated with one line of the legume, gave increased yield of total grain production.

TABLE. Grain Yield of Base Crop and Intercrops under Different Intercropping Systems

Crop/Intercrop	Yield of base crop (kg/ha)	Yield of intercrop (kg/ha)	Total yield (kg/ha)
Rice (MW 10) (pure crop)	1,563	-	1,563
Rice + mung bean (3:1)	2,511	301	2,812
Rice + groundout (3:1)	2,731	235	2,966
Rice + pigeon-pea 14:1)	2,923	433	3,356

Source: Annual Report 1980.

Where the soil is sufficiently deep, and has a good capacity for retaining moisture and nutrients to sustain the crops, a double cropping system under rain-fed conditions has been successfully adopted.

TABLE. Crop Sequence and Time of Seeding

Region	Rainy season (*kharif*) crop	Sowing time	Winter season (*rabi*) crop	Sowing time
Dehra Dun,	Maize	Mid-June	Toria	End September
Uttar Pradesh	Maize	Mid-June	Wheat	Early November
	Rice	Mid-June	Chick-pea	October
	Soybean	June	Wheat	Early November

Irrigated Uplands

Where the irrigation facility is available, the choice of the crops and cropping system from among crops suitable to the environment is governed primarily by socio-economic considerations. While in the north and north-western regions, cropping systems, like rice-wheat, maize-wheat/potato and bajra/pigeon-pea wheat, are viable, in the eastern region rice-wheat/potato, jute-rice-potato/groundnut/mung bean, and rice-potato-sesamum/mung bean are popular crop sequences. In the southern part of the country, rice-groundnut is the profitable and intensive cropping system. Wherever irrigation water is available, a third crop, such as green gram or cow-pea (fodder) in summer. is taken with profit. This is not uncommon even in the north or eastern part of the country.

In the suburban areas around cities, which have the potential for fresh vegetable sales. besides irrigation and good transport facilities, intensive sequence cropping, involving vegetables, is followed profitably. Crop sequences. like the following, are common: early cauliflower (July-October); pea (October-January); bhindi (ladies' finger) (February-June); ghindi (ladies' finger) (June-September); carrot (September-December); cauliflower (December-March); and radish (April-May).

In garden lands in Kerala, such coconut-based cropping systems as the following, have possibilities, of course, with good management: coconut + ginger/pineapple; and coconut + pepper + cocoa + pineapple.

Hill Zones

Since crop and crop varieties are thermo-sensitive, their selection for cropping systems in the hill zones must receive careful consideration. Thus, in Sikkim, up to 1,000 m above sea level, a crop rotation of maize (var. Him 123, Kissan, and Vijaya) and wheat (Girija, Shailaja. and V.L. 421) is adopted; at 1,700 m, maize-wheat/barley/potato is followed. On the other hand, a rice-wheat rotation (using IR. 8 or Pusa 33 for rice and Sonalika for wheat) has been adopted up to 1,500 m.

FERTILIZERS

The prospect of exploiting the full yield potential of crops in the rain-fed uplands is intimately related to adequate and timely cultural operations, particularly fertilizer application and weed control. For cereals, balanced fertilization with nitrogen, phosphatic and potash fertilizers is required, and for legumes, phosphatic fertilizer is essential. In arid and semi-arid zones, a moderate level of fertilizers, along with farmyard manure applied in furrows, helps good crop growth. Foliar application of urea at the reproductive phase, in some cereals (like rice and barley), increases the protein content of grains. Manual weeding is, by far, the most effective measure for weed control, where labour is adequate and available in time, but effective chemicals to control weeds can today substitute for manual weeding. Some of them are: pre-emergence application of Semazine or Atrazine at 0.5 kg a.i./ha in bajra and maize, trifluratin (0.5-1.5 kg/ha) in legumes (mung bean, black gram, cow-pea, etc.) and butachlor (2.0 kg a.i./ha) in upland rice.

DESIGN OF CROPPING SYSTEMS TO ACHIEVE NUTRITIONAL GOALS IN DRY FARMING

During the past three decades India has made rapid strides with regard to food-grain production, from about 50 million tonnes in 1950/51 to 132 million tonnes in 1980/81.

The increase in irrigated area, from 27 million hectares to 60 million hectares, on the one hand, and introduction and development of high yielding varieties and hybrids of cereals, on the other, contributed to this increase. However, the production of pulses and oilseeds remained almost static during this period. Despite the fact that rainfed lands in India contribute to more than 90 per cent of the total production of pulses and oilseeds, the dryland farmers inhabiting these lands, continue to be on the low plane of calorie and protein nutrition.

Research results from the Dryland Agricultural Research Project (All-India Coordinated Research Project for Dry/and Agriculture) show that through intercrop and double crop systems, production of pulses and oilseeds can be increased without unduly sacrificing the yields of cereals and millets. Cropping systems are now available that can meet the nutritional goals {about 10,000 calories per day for an average Indian family of five members) in dry farming areas.

Among the cereal based intercropping systems, maize + soybean supplies the highest calories and other nutrients on a per hectare basis.

However, the sorghum + pigeon-pea (2:1 row ratio) system is more stab/e over the seasons and environments and, hence, preferable. Among sequential cropping systems. the sorghum-chick-pea system holds promise of providing better nutritional standards. From the point of view of biological value and protein efficiency ratio, the rice-chick-pea sequence cropping system holds the greatest promise.

Drylands in India contribute about 75 per cent of the cultivated acreage. Even after the full irrigation potential of the county is achieved in due course, more than 50 per cent of the acreage will continue to be rain-fed. Rain-fed lands will, therefore, remain the major source for the supply of millets, pulses, and oilseeds to the growing population. Currently, these lands contribute to 80 per cent of the total production of millets and more than 90 per cent of total production of pulses and oilseeds in the country

In India, as in many Asian and African countries, the nutritional problem is one of undernutrition rather than malnutrition. It is inadequate food intake, coupled with the predominance of cereals/millets in the diet, that lead to undernutrition. Several reasons are given for the inadequate consumption of food by several population subgroups in India. Of these, lack of purchasing power, low per capita food production and availability, and large deficits in production of pulses and oilseeds that are rich sources of calories, proteins and fats, are important. Milk and meat, known for their high calorie, protein and fat contents, besides being costly, are in short supply.

Increasing and stabilizing the production of cereals/millets on the one hand, and introducing more pulses and oilseeds in the cropping systems, on the other, will go a long way towards meeting the nutritional goals of dry farming. Concerted research efforts in the last decade 11970-1980), by multidisciplinary teams engaged in dry farming research at 23 regional research stations in the country, have led to the evolution of cropping systems that hold promise, not only for high productivity, but for meeting the nutritional goals of the dry regions as well.

THE NUTRITIONAL GOALS

Nutritional goals are better appreciated on a family basis than a regional one. A farmer's family, consisting of an adult male, an adult female, an adolescent boy or a girl and two children, require a daily average of 10,600 calories, 340 9 protein, 5.0 9 calcium, and 110 mg iron. On a body-weight basis children require twice as much protein, calcium, and iron as do adults.

TABLE. Nutritional Requirement per Day of an Average Farmer's Family

Household member	No. of members	ACU[a]	Calories	Protein (g)	Calcium (g)	Iron (mg)
Adult male	1	1.0	2.800	55	1.0	25
Adult female	1	0.9	2,300	45	1 0	25
Adolescent (boy/girl)	1	1.0	2 500	80	1 0	20
Children	2	1 4	3,000	160	2 0	40
Total	5	4.3	10,600	340	5.0	110

a. Actual Consumption Unit.

CROPPING SYSTEMS FOR DRYLANDS AND THEIR PRODUCTION POTENTIAL

Cereal-based Systems

Sorghum, pearl millet, and finger millet are the major millets grown in dry regions. In regions of Telengana, Marathwada, Vidarbha, the Malwa plateau and parts of Gujarat - areas of more than 800 mm rainfall - the intercrop system of sorghum + pigeon-pea (2:1) has shown promise, with a production potential of 2.0 to 2.5 tonnes per hectare (t/ha) of sorghum grains and 0.6 to 0.8 t/ha of pigeon-pea grains, against a sole crop of either sorghum (2.5 t/ha) or pigeon-pea (0.8 t/ha). In the black soil regions of Andhra Pradesh, Karnataka, Maharashtra, and Madhya Pradesh where the rainfall is more than 1,000 mm/annum, the double crop system with sorghum, followed by chick-peas, has been found to do exceedingly well. The system is capable of giving around 3.0 to 3.5 t/ha of total productivity, in terms of grains.

In parts of Telengana, Karnataka, Maharashtra, Gujarat, Madhya Pradesh, Rajasthan, Haryana, and south Tamilnadu, which have relatively light soils, the pearl millet + pigeon-pea intercrop system has given a yield potential of 1.2 to 1.5 t/ha of pearl millet grains, and 0.6 to 0.8 t/ha of pigeon-pea grains, against a sole crop of pearl millet (2.0 t/ha). In the Jodhpur and Hissar regions, which have sierozem soils, with an annual rainfall of 350 to 400 mm, the pearl millet + green gram intercrop system has proved to be best from the point of view of total productivity per unit area.

In seasons of higher rainfall (> 500 mm) with an extended rainy season, the double crop system of pearl millet (BJ-104) followed by mustard (T. 59) can pay rich dividends in areas like Jodhpur and Hissar. The system is capable of giving a total productivity of about 3 t/ha, of which one-third is contributed by mustard, an oilseed crop. Yet another intercrop system, having finger millet as the principal crop and soybean as the companion crop, is emerging as a potentially useful system in the Bangalore and Ranchi regions.

TABLE. Finger Millet-based Cropping Systems for Drylands and Their Production Potential in a Normal Season

Cropping system	Region(s)	Grain yield (t/ha) of finger millet	Grain yield (t/ha) of soybean
Intercrop system			
Finger millet +	Bangalore	1.0-	0.2
soybean	Ranchi	2.0	0.3
Finger millet (sole)		2.5	

Maize-based cropping systems are quite popular in regions like Bangalore, Akola, Udaipur, Varanasi, Indore, Dehra Dun, Jammu, and Hoshiarpur. Intercrop systems of maize + pigeon-pea, maize + soybean and maize + green gram have exhibited distinct superiority in terms of total productivity over the other systems tried. Of these, a maize + soybean intercrop system gave the highest productivity, 2.0 to 2.2 t/ha of maize grains and 1.0 to 1.2 t/ha of soybean. Even productivity values of 3.0 to 4.0 t/ha of maize have been obtained in good rainfall years. Maize-wheat and maize-mustard double crop systems have shown great promise in the Hoshiarpur, Dehra Dun, and Udaipur regions, with a production potential of 2.5 t/ha of maize, and 2.0 t/ha of wheat in Dehra Dun, and 1.0 to 1.2 t/ha of mustard in Udaipur.

Paddy-chick-pea, paddy-finger millet and paddy-wheat double crop systems are finding favour with the dryland farmers of Dehra Dun, Varanasi, Rewa, Bhubaneswar, and Ranchi. The paddy-wheat double crop system has given, on average, 2.5 t/ha of paddy and 2.0 t/ha of wheat grains at Dehra Dun.

Pulse-based Cropping System

In areas like Anantapur and Akola, pigeon-pea forms the principal component of the intercrop system, with groundnut as the companion crop. In the Bangalore and Bhubaneswar regions, finger millet (transplanted) forms a good companion crop to the pigeon-pea, the yield of finger millet being obtained as a bonus. Among double crop systems, green gram-safflower, cow-pea-finger millet/maize, black gram-rabi sorghum systems have shown great promise, cow-pea as the principal crop being quite promising in the Bangalore region.

TABLE. Pulse-based Cropping Systems for Drylands and Their Production Potential in a Normal Season

Cropping system	Region	Principal/First crop yield (t/ha)		Companion/Second crop yield (t/ha)	
		Grain	Edible portion	Grain	Edible portion
Intercrop system					
Pigeon-pea + groundnut	Anantapur, end Akola	1.2- 1.5	1.0- 1.2	1.0 1.2	
Pigeon-pea + finger millet (transplanted)	Bangalore, and Bhubaneswar	1.0 1.5	0.8- 1.2	0.8 1.0	0.8 1.0
Pigeon-pea (sole)		1.5	1.2		
Double crop system					
Green gram - safflower	Bijapur, Solapur, Indore, Udaipur, Ranchi, and Hissar	0.6- 0.8	0.5 0.7	1.0 1.5	
Cow-pea - finger millet	Bangalore	0.6- 0.8	0.6- 0.8	2.0 2.4	2.0 2.4
Cow-pea - maize	Bangalore	0.6- 0.8	0.6- 0.8	2.2- 2.5	2.2 2.5
Black gram - rabi sorghum	Bijapur, Solapur, Akola, and Indore	0.5- 0.6	0.4- 0.5	1.2- 1.5	1.2 1.5

Oilseed-based Cropping Systems

TABLE. Oilseed-based Intercropping Systems for Drylands and Their Production Potential in a Normal Season

Intercrop system	Region	Yield (t/ha) of groundnut	Yield (t/ha) of the companion crop
Groundnut + red gram	Anantapur, Solapur, Akola, Ranchi, and Rewa	0.9- 1.0	0 5 0.6
Groundnut + castor	Rajkot, Ranchi, and Bangalore	2 0- 2.5	0.6 0.8
Groundnut (sole)		2.5	

Red gram and castor go very well with groundnuts as the principal crop in the intercrop systems, the former (groundnut

+ red gram) having shown promise in the Anantapur, Solapur, Akola, Ranchi, and Rewa regions. In the Rajkot, Ranchi, and Bangalore regions, groundnut + castor system is becoming popular.

TOTAL NUTRIENTS SUPPLIED THROUGH CROPPING SYSTEMS

Cereal-based Systems

Of the cereal-based intercrop systems, the maize + soybean system supplies the highest quantity of various nutrients - calories, proteins, fat, calcium, iron, etc. - on a per hectare basis, followed by the sorghum + pigeon-pea (2 :1) system. Between pearl millet + pigeonpea and pearl millet + green gram, the former supplies more nutrients per unit area than the latter. Pearl millet-based systems are, however, richer in iron than either sorghum or maize-based intercrop systems. In terms of calcium and iron, finger millet-based cropping systems are the richest .

Pulse-based Cropping Systems

The pigeon-pea + groundnut system supplies the highest calories, proteins, and fat per hectare. However, the highest calcium and iron are supplied by a pigeon-pea + finger millet system .

Double Cropping Systems

The maize-wheat sequence crop system supplies the highest calories and proteins, followed by the sorghum-chick-pea/rice-chick-pea systems. The sorghum-chick-pea system supplies, in addition. abundant calcium and iron.

NUTRITIVE VALUE OF THE PROTEINS OF SOME FOODSTUFFS

The biological value (BV) and protein efficiency ratio (PER) of some of the more frequently consumed foodstuffs are presented in table below. Among cereals (rice, wheat, and maize), rice has the highest BV and PER. Between the two pulses. chick-pea and pigeon-pea, the chick-pea proteins have a higher nutritive value than the latter. Regarding edible oil sources, sesame is superior to groundnut. The BV and PER of some animal/poultry/fish products are also presented in the table for comparison with plant sources.

TABLE. Nutritive Value of the Proteins of Some Foodstuffs

Foodstuff	Biological value	Protein efficiency ratio
Rice	68	2.2
Wheat	65	1.5
Maize	59	1.2
Chick-pea	68	1.7
Pigeon-pea	57	1.5
Groundnut	55	1.7
Sesame	62	1.8
Egg	94	3.9
Milk	84	3.1
Meat	74	2.3
Fish	76	3.5

Source: McDivitt and Mudambi 1969.

CONCLUSION

From the various tables presented in the paper, it is obvious that a decade of research on dryland agriculture has made available the technology to produce enough nutrition, through plant sources, for five members of a marginal/small farmer's family. This technology is, essentially, based on proven intercrop/double-crop systems suitable to different agro-climatic regions. Of course, inputs such as improved varieties, judicious fertilizer use and timely plant protection measures, including weed control, are essential to optimize the yield levels in a normal season.

DISCUSSION

Agricultural production and nutrition

The Green Revolution has set in motion sweeping changes in food-growing patterns, and the nutritional quality of the food mix cannot be taken for granted. An ICRISAT study has shown that protein availability has gone up substantially in the wheat production system. Both nutrient and production cost data for other systems and regions of the country must be gathered and compiled for the benefit of those planning food systems. Since improved protein quality is derived when the diet is a cereal-pulse blend, the total system has to be considered. New data led a recent FAO/WHO/UNU consultation to raise the adult protein requirement from 0.57 to 0.72 grams/kg body-weight, and this again involves the provision of proteins of better quality by food blending.

In farm practice, anti-nutrients are a factor to be taken into account. Lathyrus is an excellent example. It is an exceptionally hardy. easy-to-grow crop, favoured by marginal farmers in India and Bangladesh. It is raised as a mixed crop with barley, wheat, and chick-pea in many parts of India, and as a pure crop in Madhya Pradesh and even Andhra Pradesh. Screening of Lathyrus varieties in Bangladesh revealed that 5 to 7 per cent had a very low neurotoxin B-oxalyl aspartic acid content, and a Pusa variety totally devoid of the toxin had been evolved. Some attention is called for to pocket areas that employ unusual food materials in fair quantity. Thus, Lathyrus, chinnapodium, jackfruit, and zizyphus are important in various local regions of Bangladesh. Whether the trypsin-inhibitors found in most pulses are relevant in human nutrition is still in doubt, though they are of course destroyed by cooking. Correlation in chickpeas between methionine and total sulphur content is poor, but seed coat smoothness and polyphenol content do correlate with insect infestation by foiling oviposition.

Consumer Acceptability

The overriding emphasis in agricultural development on high productivity per unit area has tended to ignore the preferences of the consumer. It is true that unacceptable

Mexican red wheats of the early 1960s were rapidly phased out, but it has taken longer for the great variability in size and shape of JR-8 rice grown in different regions to be recognized, besides its low milling yield. The same variability is noted in the length/breadth ratio of Hamsa rice, which is an important index of consumer acceptability.

A way out of the rice classification problem for market purposes is to also take into account thickness, which seems to be dependent on nitrogen fertilization of the crop. The Phalquna rice variety has rapidly become extremely popular in the Andhra coastal belt. The black tip, which caused considerable alarm, has fortunately been found to have no adverse nutritional consequences, though it does of course affect consumer acceptability. A similar problem with regard to wheat is Karnal bunt which particularly attacks WL-711 wheat grown in Punjab and Haryana; the Government has had difficulty in fixing specifications, and consumers are apprehensive. High-yielding hybrid jowar may give good returns to the farmer, but is totally unacceptable to consumers in Maharashtra for its off-odour and peculiar appearance. Again, while a maize-soya rotation mentioned in one of the papers may be excellent in terms of yields and economics, consumer demands for both these crops are minimal.

ECONOMICS AND MARKETING

While the pattern of urban marketing has stabilized to a certain extent, rural marketing is a neglected area. Even such foods as milk and tomatoes fetch low returns for the actual producers, and poverty among them is increasing. The widely dispersed public distribution system through fair-price shops must be widened to include these commodities. In fact, stability of farming systems both in terms of crops raised and of remunerative prices for them would provide employment opportunities, which are the crying national need. When the productivity of natural resources of land and water is maximized, and the farmer is provided with options which are stable and economically attractive, he/she will make the right crop-raising decisions. Public policy must help the farmer realize these expectations. So far. only cereals have offered really viable technologies.

6

Improved Post-harvest Technology to Maximize Yield and Minimize Quantitative and Qualitative Losses

The need for improving the high yielding varieties and also the application of fertilizers, irrigation water coupled with multiple cropping systems appeared a far more attractive approach to achieve higher yields during the 1970s. Hence, not enough attention was given to the post-harvest operations. But the limitations of the pre-harvest operations became quite evident in a short span of time resulting in that of an appropriate attention towards the reduction of losses in the post-harvest operations. The postharvest losses of food grains in developing countries are enormous. When the consequences of such losses are evaluated in terms of human sufferings and economic cost, they represent an international challenge. Successful prevention of these losses can be a tremendous help to the human race all over the world. The quantum of losses vary from country to country and also in the same country for different crops due to variations in environmental factors and human efforts utilized for prevention. Under Indian conditions, the post-harvest losses amount to about 10 per cent of the total grain yield. In a systematic study carried out on paddy crops, the losses appeared to be contributed by the following major post-harvest factors. (a) harvesting, 1 to 5 per cent; (b) transportation, 2 to 7 per cent; (c) threshing, 2 to 6 per cent; (d) storage, 2 to 6 per cent; (e) drying. 1 to 6 per cent; and (f) milling, 2 to 10 per cent.

Thus, these losses cumulatively could amount to as high as 40 per cent by weight and are normally accompanied by loss of quality also. Quality loss of grains can take place at all stages but more severely at the stage of drying, storage, and milling. These factors indicate the urgency of attaching the highest importance to the management of post-harvest problems, if the gains of higher production are to be fully exploited for the benefit of the community and the country.

Quality prevention during post-harvest operation in general and the same during storage in particular is also important for preserving the nutritive value of food grains. The grains undergo some chemical changes that alter the flavour and nutritive values.

The insect and fungi not only consume the grain but also accelerate the harmful chemical changes. Some fungi produce poisonous substance or toxins in grains. Aflatoxins are the most dangerous toxins produced by Aspergillus.

The quantitative losses can be caused by mechanical agents such as birds, animals, hailstorms, rains, over drying, and shattering in the fields during harvesting. Biological agents such as birds, rodents, mites, and insects are believed to be the most destructive agents of grains and seeds in the stores. The losses occurring due to other factors such as:

(a) changes in moisture content;
(b) dust and broken grains;
(c) reduction in germination power,
(d) loss of palatibility;
(e) heating and caking, etc. can render the grains unacceptable for use as human food or as animal feed.

The results obtained from various studies presented in the paper support the idea that substantial increase in nutritional food availability can be achieved through postharvest loss prevention at farm level as well as storage points in trade system.

INTRODUCTION

Developing high-yielding varieties was definitely the most attractive route to higher production in the early 1970s. Not enough attention was attached to post-harvest operations, which were considered less significant. However, the limitations of pre-harvest systems became quite evident shortly thereafter. Staggering losses to the extent of 40 per cent or more were reported in many countries, and the importance of preventing post-harvest losses are now fully recognized.

Post-harvest losses were long considered to be contributed by storage alone, and not much attention was paid to losses occurring during harvesting, transport, drying, threshing, milling and pre-milling treatments. Quantitative loss of food grains was the main concern of the producers and the consumers. Qualitative losses at various stages were not fully recognized, but systematic studies by different organizations revealed that these losses are enormous and should be prevented.

Thus in paddy crops, the following major factors contribute to post-harvest losses (Ojha 1978b): harvesting, 1 to 5 per cent; drying, 1 to 6 per cent; transportation, 2 to 7 per cent; storage, 2 to 5 per cent; threshing, 2 to 6 per cent; milling and pre-milling, 2 to 10 per cent. These losses cumulatively could be as high as 40 per cent and are normally accompanied by loss of quality also. Quality loss of products can take place at all stages but more particularly during drying or curing of harvested crops, storage, and milling and pre-milling treatments. Post-harvest attention is essential if the grains of higher production are to be fully exploited. Also the quality loss occurring in food grains can be extremely serious if the farming community is not aware of them. To illustrate the point, freshly harvested high-moisture paddy, bajra, or groundnut crops, if not dried to a safe moisture level, can become poisonous by the development of toxins through mould growth.

HARVESTING AND THRESHING OPERATIONS

Harvesting is a short-period operation and a labour-intensive one. It takes about 170 to 200 person hours to harvest one hectare of paddy field. If harvesting is not initiated at an appropriate moisture content, the matured paddy or wheat grains are likely to shatter in the field or during subsequent movement to the threshing years. Such losses may be as high as 25 per cent depending on variety and stage of harvesting.

Harvesting Stage for Quality and Quantity

Farmers in general are not well aware of the advantages of early harvesting of field crops. Conventionally, most farmers in the mono-culture system of paddy cultivation harvest the paddy crop at about 16 per cent moisture content, and suffer a loss of 10 to 15 per cent of the expected field yield. A definite correlation between grain moisture and days after flowering has been established through field studies (Bal and Ojha 1977). About 0.36 per cent moisture was observed to be lost every day from the Jaya variety between the fifteenth and forty-fifth day after flowering. Normally about a week would pass during which the grains are at a moisture content between 20 and 24 per cent. If the crop is harvested during that period, the losses are considerably reduced and the field yields are comparatively high. Maximum yields were actually recorded from the harvest carried out on the thirty-first and thirty-second day after flowering, all field yields being corrected to a fixed moisture content of 14 per cent. An appropriate period for harvesting paddy crops may therefore lie between the twenty-eighth and thirty-sixth day after flowering. This provides about nine days to complete the operation without sacrifice of either the yield or quality of the grain. If the crop is over-dried, sun-cracks develop on the kernel and cause breakage during threshing and milling.

Variation in Milling Quality

Milling quality of paddy is defined by the terms "total yield" and "head yield" (yield of full-size rice), both expressed as percentage weight of the original paddy. The total yield of the Jaya variety increased from 54 to 73 per cent between the sixteenth and thirtieth day after flowering. The poor total yield on early harvesting was a result of immature grains on the panicles. The total yield beyond the thirty-sixth day after flowering again showed a reducing trend because of cracks which had developed in the kernels. Thus study again supports the contention that the paddy crop should be harvested within 28 to 36 days after flowering. This would be the optimum harvest period for most high-yielding paddy varieties of India. Pillaiyar (1977) reported that the maximum yield could be obtained by harvesting paddy crops 15 days after ripening of the tip grain of the primary earhead. According to Khan (1976), harvesting should be carried out when the hulled grains on the upper portion of the panicles are clean and firm, when the grains at the base of the panicles would be in the hand dough stage.

Grain damage can be caused not only by delayed harvesting but also by improper handling subsequent to harvesting. Devnani (1976) and Chabra and Singh (1977) reported that over-dried grains are prone to fracture during threshing, whereas moist grains are likely to bruise through improper cylinder-concave clearance and high thresher cylinder speed. If not carried out properly, harvesting and threshing by machines can be more injurious to seeds than manual methods. Of course, threshing by beating with sticks can crack or break the grains into pieces and should be discouraged as a means of seed production.

DRYING AND CURING OPERATIONS

The desirable moisture content at harvest appears to be between 20 and 24 per cent for paddy and 15 to 20 per cent for wheat. As the grain has to be stored for an extended period, it must be dried either in the sun or in a mechanical dryer utilizing heated air.

Harvesting and drying during rainy weather in the coastal areas, and in the eastern regions of high precipitation, have always posed a serious problem of quality deterioration in paddy crops. Mechanical drying is the only acceptable solution.

The continuous spell of humid and rainy weather prevailing during the harvesting period of the aus or ahu varieties of the north-east region and the kuruvai varieties of the Cauvery delta conduces to the growth and proliferation of microbes on paddy grains. Subrahmanyan (personal communication 1978) reported that the higher levels of micro-organisms observed in the flowering and milk stages of paddy might be due to the presence of a higher level of readily available nutrients in free form, like reducing sugars and amino acids. The chances of increased infestation and rapid multiplication, leading to heating and discoloration, become more acute where immature grains are not separated after threshing. Those organisms multiply rapidly when grains are moist and germinate. Such fungal forms as Mucor, Aspergillus flavus, A. niger, A. fumigatus and Penicillium are generally observed in paddy harvested in the wet season. The vegetative growth of these fungi ramify the grains and cause webbing. Fungi webbing and grain germination result in caking when moist paddy is bagged in gunny bags.

Paddy Drying

Recognizing the need for better drying facilities for both freshly harvested and parboiled paddy, mechanical drying systems were introduced in the late 1960s in India. These dryers were of Louisiana State University (LSU) design, with a holding capacity of 6 tonnes or more. As the heat-generating fuel was furnace oil, these dryers tended to be uneconomical. Later, a few portable dryers of small holding capacity that employed rice husk as fuel were developed and are being used commercially (Ojha 1978a). Among these, the RPEC recirculatory dryer and LSU dryer coupled with a husk-fired furnace gave good performance (see FIG. 2a. Pictorial View of Recirculating Batch Dryer Coupled with Husk-fired Furnace for Paddy Drying (Rice Process Engineering Centre (ITT), Kharagpur)). Bose et al. (1980) developed a system utilizing a fixed-bed LSU dryer coupled with a solar-cum-husk fired heating unit for drying high-moisture paddy. Since these dryers were of about one tonne holding capacity, they were adopted by smaller millers and others.

The circulatory dryers can be successfully used to dry high-moisture (24 per cent moisture content) paddy to about 16 per cent m.c. in four passes of 30 minutes each. Paddy so dried gave a high head yield due to uniform drying of individual grains. Table 1 presents yield data after drying at various temperatures in mechanical dryers (Rama Rao 1974). Drying of high-moisture grain, either for seed or for human food, is essential. A typical cross-flow recirculatory drying system is shown in figure 2b. Grain flows from the wet grain holding bin at the top, down the columns and is discharged at the bottom. The column of grain is held between perforated sheets and air passes through the column perpendicular to the grain flow. In one design, the entire column receives the heated air for removal of moisture, whereas in another, the upper portion of the columns are drying sections, and in the lower portion the grain is cooled. The flow rate is regulated by a metering device at the outlet of the column. The metering device can be either automatically or manually controlled depending upon the particular requirement.

TABLE. Performance Data for Drying Parboiled Paddy with Recirculatory Batch Dryer without Tempering

Drying air temperature (°C)	100	90	80	70	60
Average total yield (%)	72.8	72.7	72.7	73.0	72.3
Average head yield (%)	52.0	50.0	55.0	58.5	61.6

Source: Rama Rao 1974.

A multipass system is used in India for drying. Each pass takes about 30 minutes with the temperature of the heated air being 50°C for raw paddy and as high as 90°C for parboiled paddy. About 2 per cent loss of moisture occurs in each pass. Moisture removal can be accelerated by tempering the grain in a heap or in a silo at the end of each pass. The most common cause of cracking is believed to be either rapid drying or rapid moistening, or both. Cracks are a result of stresses developed in the grains, and they are more frequent in kernels of large sizes.

PRE-MILLING TREATMENTS

Soaking in cold water and drying in sun, soaking in hot water followed by sun drying or steaming, soaking, steaming, and drying are the pre-milling combinations practised in India to improve the milling, nutritional, and keeping qualities of rice. Some varieties of paddy possess poor milling quality for many reasons.

The weather prevalent during field maturity and harvesting, or unscientific post-harvest practices in handling and storage, render the paddy grains fragile and lead to breakage during milling. Proper hydrothermic treatments, as earlier described, can improve the milling qualities of paddy. One such treatment is parboiling, which is given to about 60 per cent of all the paddy produced in India. Improved methods of parboiling include the three steps of hot-soaking, steaming, and drying. Parboiled paddy is produced either as a small scale, family trade, or at a commercial level in rice mills. It would not be feasible to discuss here the various parboiling methods practised in India. However, the quantitative and qualitative improvements rendered by parboiling will now be mentioned.

Nutritive Value of Parboiled Rico

The rice grain is composed of a surface layer called bran (itself consisting of the pericarp and aleurone layers), the starchy endosperm, and the germ. The surface layers are rich in protein, oils, and vitamins compared to the endosperm. Parboiling augments the nutritive value, through first translocating vitamins from the peripheral layers to the central core, and oil from the endosperm to the bran layers, followed by heat sealing them.

Milling Quality of Parboiled Rico

It is well known that parboiling reduces breakage during milling. Thus improvement is usually attributed to increased hardness, or the healing of sun-cracks and other defects resulting from biochemical changes. The percentage of brokers is drastically reduced and the head rice outturn is remarkably increased. Parboiling leads to an increase in total yield as well as head yield. The performance study report on rice mills reports an increase of about 6.6 per cent in total yield and over 15 per cent in head yield for a few varieties (Mathrani 1971). Varieties that give poor yields in raw milling often show excellent milling quality after parboiling.

TABLE. Head and Total Yields or Raw and Parboiled Rice of Mixed Varieties

Type of rice	Total yield (%)	Head yield (%)	Remarks
Raw	71.3	52.1	Modern mill of the Dandekar
Parboiled	73.0	53.9	type was used for milling.

Source: Mathrani 1971

Bran from parboiled rice contains less starch and more oil than raw rice bran. The oil content of parboiled bran varies from 15 to 25 per cent, about 5 per cent higher than in raw bran.

Cooking Quality of Parboiled Rice

The cooking quality of rice may be defined as the ability of the grain to be cooked to the required texture or tenderness without losing shape. A grain that cooks to a fluffy, non-cohesive, tender texture, and retains its shape, while increasing in size, is said to have good cooking quality.

Losses in Weight during Perboiling and Drying

Large-scale field experiments carried out on commercial plants have indicated that weight losses during parboiling and drying varied from 0.85 to 1.75 per cent of the total input. The lowest percentage losses were observed with the modern parboiling methods, such as the CFTRI and pressure parboiling methods in which hot soaking of paddy is completed in less than four hours.

MILLING OF GRAIN

The term milling generally refers to the size-reduction of granular material, but for food grains the term has different connotations. Wheat milling means wheat grinding to prepare flour, maida, suji, etc. Rice milling includes operations like dehusking and polishing; dhal milling also involves processing operations such as husk separation, splitting of kernels and polishing, or polishing only.

The traditional methods employed for milling are wasteful, non-uniform, and do not yield useful byproducts. Modern and the traditional methods can be compared using the outputs obtained, which are shown in table 3. About 7 per cent extra yield can be derived by employing modern rice milling machines. On an all-India basis, an additional quantity of 5 million tonnes of rice and Rs 6,000 million worth of pure rice bran can be achieved. A better quality of rice is produced through modern mills. The degree of polish and the percentage of brokers in the final product can be very precisely controlled to suit the specific needs either of industry or of the consumer. The most impressive gain that rice mill modernization has brought about is the efficient utilization of by-products of rice. If the total quantity of available bran is solventextracted, it will yield about 800,000 tonnes of bran oil for the country.

TRANSPORTATION LOSSES

Movement of harvested field crops to the threshing yards, and of the threshed grains to various storage and consumption points, results in a partial loss of food grains through handling, shipments, pilferage, predation by birds and rodents, and weather hazards. Conservative estimates of such losses are noted in the figure. Most of these losses occur through careless handling of gunny bags by the labourers or porters, and leaky bags and containers employed by user agencies. Movement of food grains from warehouses of the Food Corporation of India (FCI) or the Central Warehousing Corporate (CWC) using road transport frequently results in considerable spoilage of grains on the roadside. Food grains being moved from Punjab to West Bengal, a distance of 1,500 km, in open rail-wagons frequently are subject to heavy showers on the way, causing partial or total spoilage through caking, discoloration, and sprouting.

Loss of food grains from railway yards through birds. rodents, and miscreants is a common occurrence in commercial centres and metropolitan towns. Singh (1974) indicated the following range of transit losses for food grains in various modes of transport: rail, 0.5 to 3.5 per cent; road, 0.25 to 1.25 per cent; and river, 0.5 to 1.0 per cent. In 1970/71 the losses of food grains in medium to long-distance rail transit were estimated by FCI at about 1 per cent. Safe, and less-wasteful, transport is needed for movement of food grains.

GRAIN STORAGE AND PRESERVATION

In India about 30 to 40 per cent of the grain produced in the country is handled by the Government and private agencies, while the remaining 60 to 70 per cent is retained by village farmers for household use and consumption. The storage and warehousing facilities built by the FCI and CWC were located, mostly, in or near urban centres. Storage at the village level was neglected until village co-operatives initiated the building of improved storage facilites with the financial support of the National Co-operative Development Corporation (NCDC). though the total storage facilities thus provided by the co-operatives is very meagre (Prasad 1980). The loss of food grains stored in FCI and CWC godowns is less than 1 per cent, but in the village storage systems it can be as high as 5 per cent as a national average. Wimberly (1972) placed storage losses at 5 to 10 per cent. Undoubtedly there is a high loss each year that is still to be controlled.

FARM-LEVEL STORAGE

Public agencies handle only about 10 per cent of the total produce. The other 90 per cent is handled by farmers and the private trade both for seed and family consumption purposes. Even a conservative 3 per cent loss in farm storage would work out to the loss of 4 million tonnes of food grain, annually. However, a study by Khare et al. (1972) indicated maximum and minimum weight losses of 4.5 and 1.7 per cent, depending on the store type and climate. In a similar study on wheat storage in villages around Delhi, Zutshi (1976) noted that farmers suffered a weight loss of only 0.8 per cent by storing in earthen pots as against 15 per cent in bulk in a room.

Quality Control and Nutrition

Prasad (1980) points out that good-quality grains have a natural attraction for the consumer. Quality depends on storage, and includes the nature of grain, period of storage and method of preservation. Storability differs from grain to grain. All types of grain deteriorate through the action of moisture and pests. The effects of pests are relatively greater in raw rice than in parboiled rice. Paddy is less damaged by pests than is rice. Of course, though the keeping quality of wheat is better than that of rice, even wheat products have a shelf life of not more than a few weeks. Pulses are prone to greater insect damage in storage than cereals, and split pulses or dhals can be stored longer than the whole materials. Coarse grains such as maize, bajra, and jowar are softer in texture than rice and wheat, and, therefore, relatively more susceptible to damage by pests as well as moisture. They also tend to become rancid after 18 to 24 months because of a high-percentage oil content.

Stored grains undergo chemical changes that alter their flavour and nutritive value. Insects and fungi not only consume the grain but can also accelerate harmful chemical changes. Some fungi produce poisonous toxins in food grains. The discovery of the poisonous metabolite aflatoxin in groundnuts brought to light the hazards of consumption of mouldy food material. Insect infestation causes loss of carbohydrates and thiamin. Fats present in food grains tend to break down during improper storage, and this is accelerated by insects, mould attack, high moisture and temperature. Breakdown of fats causes an increase in free fatty acids, and oxidation brings on rancid flavours. Infested food material becomes unpalatable through the presence of excrete and live or dead insects, and even results in poor growth of the body.

There is a very close relationship between the quality maintenance and the nutritive value of food grain. Quality is manifested in many aspects, such as appearance, freshness, mustiness, insect or fungus attack, moisture content, freedom from foreign matter, pesticide residues, filth, uric acid content, and hidden infestation.

7

Interaction at the Post-harvest Stage

The outer bran layers, in coarse grains, are fibrous and contain tannins, pigments, and bitter principles. Their removal would considerably enhance their palatability and may even add to nutrition. Appropriate techniques for effecting this have been developed.

For grains with hard endosperm (sorghum, maize, and pearl millet) incipient wet conditioning with 2-3 per cent moisture followed by abrasive milling could remove the outer bran. Polishing should be judicious as over-polishing causes undue depletion of nutrients. The polished or pearled grains could be used as whole grain, semolina, or flour Alternatively, similar wet conditioning, followed by immediate grinding in a plate grinder and sieving, gave a highly nutritious flour. nearly free from bran.

Heat treatment of these grams was found to affect their nutritional, physical, textural, and storage properties. Steam treatment of ragi reduced the stickiness of the cooked mudde or dumpling and contributed grainy texture to it. Heat treatment reduced the hot paste viscosity of millet flours, inactivated the lipase, and increased the shelf-life of whole-meal flours. Drastic heat treatment, however, triggered development of oxidative rancidity. Puffed or popped coarse grains are also highly acceptable products.

Elaboration of amylases during germination was faster in pearl millet and ragi than m sorghum or maize. Amylases degraded the starch, reduced its hot paste viscosity and increased the caloric density of weaning foods based on malted grains. Weaning foods based on malted ragi, millet, sorghum, and maize, and with a legume supplement, have been developed. Other advantages of germination are increase in lysine, hydrolysis of phytin, and release of vitamin C. Varietal differences in respect of amylase activity during malting have been noted.

Semolina from sorghum and maize could replace rice or wheat semolina in idli and dose. Lactic fermentation of gelatinized ragi or other cereal flours has recently been studied.

Some form of processing usually precedes the consumption of all food grains. The processing treatment could be milling to remove bran/husk, heat treatment to cook it and remove raw flavour or taste, puffing, popping, or roasting to introduce crunchy texture and impart desirable aroma, malting, or fermentation to improve nutrition or digestibility, canning for convenience, dehydration for enhancing storage life, and so on. Some forms of processing or over-processing may also adversely affect the nutritional qualities of the food grains.

An enormous amount of work has been carried out and documented with respect to the science and technology of such processing and its effects on the nutritive value. culinary quality, storage life, and utilization potential of food grains, particularly of such major cereals as wheat and rice and the majority of legumes and oilseeds. Generally, optimal processing has been found to be necessary and usually beneficial, although severe or long processing adversely affects nutritive value, storage life and their culinary quality. For instance, while milling is necessary, over-milling or a high degree of extraction reduces yield and nutritive value. Optimal heat treatment improves flavour, texture and nutritive value, but severe heat treatment may destroy vitamins, bind nutrients or even initiate rancidity. The processing should therefore be judicious or optimized to derive the best utilization potential from it.

Such information on coarse grains like sorghum, maize, and other millets is recent and quite meagre. The object of the present paper is to highlight and review some of our recent work on these grains, which happen to be staples consumed by large sections of the population in the relatively drier and semi-arid parts of Asia, Africa, and South America. The processing treatments that will be considered are milling, heat treatments, germination, and fermentation.

MILLING

The outer bran in coarse grains is fibrous, bitter, astringent, or coloured. Milling of the coarse grains is therefore desirable to confer adequate consumer acceptability to them. The technique and the machinery for the purpose have been standardized. The grains are wetted or moistened with 2-3 per cent moisture, conditioned for 5 to 15 minutes, and the bran removed by an abrasive device used in rice milling (Viraktamath et al. 1971). It is desirable to regulate the milling in such a way that only the outermost bran layers are removed retaining the germ and the nutritive aleurone layers intact. The polished grains could be used as such, or ground to flour or semolina for culinary uses (Desikachar 1975).

Where the grain being processed is soft or small-sized, the moist conditioned grain is ground in a plate grinder mill (chakki. plate or emery grinder) and immediately sieved. The bran, which resists grinding to fine particles, is removed on a 16-20 BS mesh sieve. The flour or semolina obtained is a fairly refined product with minimum bran contamination (Desikachar et al. 1980).

It is obvious that overmilling or very high refining must be avoided, since it removes the aleurone layers and germ rich in protein, vitamins, and minerals. The chemical composition of the fractions at a different stage of polishing sorghum illustrates this point. A 85-90 per cent extraction flour or a 10-15 per cent polished grain retains a high proportion of nutrients with minimum bran contamination (Raghavendra Rao and Desikachar 1964).

HEAT TREATMENT

The presence of a dense horny endosperm in sorghum and maize hinders quick or soft cooking of the millet grains. They are therefore used as flour for making roti, chapati, etc. Size-reduction of grain into semolina, or flattening it to a flake, enable quicker cooking (Raghavendra Rao et al. 1979). Sorghum semolina is found to be highly resistant to quick hydration. Cooking, especially in an acid medium, improves digestibility.

The effects of pre-cooking, steaming, roasting, puffing, and so on are generally beneficial. For example, the slimy character of starch in its cooked form is reduced by prior heating of the grain (Desikachar 1974). Cold slurry viscosity is increased and hot paste viscosity reduced by heat treatment (Raghavendra Rao et al. 1981). The storage life of the whole-meal flour/semolina is enhanced by heat treatment, since the lipase is inactivated and hydrolytic rancidity eliminated (Wamajee 1978; CFTRI Report 1976). Severe heat treatment, however, sets off oxidative rancidity.

Popped or flaked millet grains have been made use of for development of different types of weaning foods or of supplementary foods for feeding school and pre-school children (Desikachar et al. 1981). Popping or puffing is especially useful here, because it imparts a highly crunchy texture and desirable flavour. The millet is supplemented with pulses to increase protein quality. Varieties suitable for popping have been identified in sorghum and ragi (Viraktamath et al. 1972; Malleshi and Desikachar 1981a).

GERMINATION

Malting of grains, particularly pulses, is a traditional practice in certain parts of India. The benefits of germination are many. Vitamin C is elaborated, phosphorus availability is increased, and lysine and tryptophan are synthesized (Dulby and Tsai 1976). Both starch and protein are partially degraded making for better digestibility (Ram et al. 1979). Some of the flatus factors are also degraded (Rao and Belavady 1978). The elaboration of amylases results in a lower viscosity and thinning down of the starch slurry (Brandtzaeg et al. 1981). There is also an overall improvement in the flavour profile. In view of these multiple benefits. use of germinated grains, to a larger extent, should be advocated. The germination, however, should not be prolonged as otherwise too much root formation and metabolic weight loss of grain may become a problem.

Elaboration of amylases during malting has been taken advantage of in the development of weaning food formulations for babies, wherein a low hot paste viscosity and a high calorie density is desired. One formulation consists of malted ragi and green gram with supplemental vitamins and minerals. The formulation has been widely tested and found to have good consumer acceptance. Its nutritional value has been found to be comparable to that of common proprietary weaning foods. Addition of a small amount of malt flour has been found to bring down the paste viscosity of many weaning foods (Malleshi and Desikachar 1981b).

FERMENTATION

Fermentation of foods has also been practised for improving the flavour, texture, and palatability of foods. Bacterial or yeast fermentations are more common in India, while fungal fermentation is more widely practised in the Far East. Synthesis of B-complex vitamins, partial digestion of starch and proteins, and release of enzyme are brought about during fermentation (Ready et al. 1980), Increase of vitamin B12 and folic acid, and the synthesis also of a few antibiotics, have been noted, particularly in fungal fermentation used to make "tempeh" and "natto" (Van Veen and Steinkraus 1970). The overall effects of fermentation are highly beneficial. Millet grains can substitute for the rice or wheat component of such fermented foods. Maize or sorghum semolina can replace rice or wheat semolina in making idli (Cheeptongkum 1976). Sorghum. maize or ragi can replace rice in dose formulations (Raghavendra Rao et al. 1979). Sorghum, millet, or maize flour can also replace up to about 20 per cent of the wheat used in making bread (Haridas Rao and Shurpalekar 1976).

NUTRITIONAL IMPLICATIONS OF RECENTLY DEVELOPED TECHNIQUES OF STORAGE AND PEST CONTROL

Millets are nutritionally deficient compared to cereals. Root crops are also similar to millets. Insects, moulds, and mites that attack these commodities and bring about big-deterioration, even at comparatively low moisture or water activity, are responsible for producing in situ undesirable metabolities, apart from depletion of calorie and selective nutrilites. Enzyme of seeds are activated under ecosystems of underground pit storage and above-ground silo storage systems. Changes in bulk densities, production of frass, microbial and entomogenous debris, related mycotoxins and entomotoxins are responsible for the lowering of nutritional quality besides posing health hazards to animals and man.

Prevention of such depredations by pests and big-deterioration can lead to predictable nutritional and energy supplies. In tropical and subtropical ecosystems, pre-harvest prophylaxis to ensure post-harvest nutrition availability is a primary requirement. It is pre-emptory to apply selectively deterrent pesticides for annhilating fungi and insect pests, and repellents for birds and rodents, to protect the ripe grain panicles. This will reduce the gap between the predicted harvest forecast and actual harvest.

The post-harvest threshold as a pea of the system of crop production must be treated in conjunction with prophylactic and curative approaches, as pre-harvest internal infestations are non-retrievable episodes that nullify efforts to achieve nutritional self-sufficiencies or surplus food supplies.

Appropriate, safe (free from health or environmental hazards). and tropically matching technologies available in the world are catalogued for adoption under the agro-climatic and socio-ecosystems. Preharvest prophylaxis; solar and crib drying; thermally compatible storage structures, non-toxic seed or raw grain protectants, Meta-H-Halloysite; nutritional grain/ processed product protectants; small-scale and large-scale disinfestation technologies are stated.

FOOD AVAILABILITY VERSUS THE PEST COMPLEX

All agricultural products are perishable. Cereals, pulses, and oilseeds are relatively durable commodities, in contrast to fruits and vegetables and other green agricultural products. Insects, moulds, mites, and rodents attack the commodity in the field and in the post-harvest period (Cotton 1963). Their depredations not only deplete quantity, but also damage the quality of the commodities. More insidious effects of the big-deterioration caused by insects, moulds, and mites are reflected in the production of entomotoxins and mycotoxins (Majumder 1975, 1978). Crops like sorghum, bajra, ragi, pulses and beans, maize, groundnut, and many others are attacked by field pests and field fungi. They directly bring about shedding and, hence, shattering of grains during the harvesting operations. Yields of crops are, therefore, affected very adversely, and discrepancies arise between production forecasts and the quantity actually harvested. When infestation is high (at the pre-harvest stage), post-harvest quality is proportionately reduced and shelf-life is also lowered. Thus, on the one hand, a direct reduction in yield occurs, and, on the other hand, the harvested quality suffers.

Agro-climatic situations also affect post-harvest quality and the storage characteristics of food grains and fruits and vegetables. The higher the thermal and humidity effects of the environment during the production of the crop, the lower will be the resistance to infection and infestation during the postharvest period. Drying in the field in the post-harvest period poses difficult problems, as the water activity and consequent enzymatic activities within the tissues remain at higher potentials. Such postharvest practices as underground pit storage for millets and related crops, or above-ground storage of other cereals in bins and silos, activate the enzymes of the seeds and also accelerate moisture migration, together with thermogenesis (Majumder 1972).

Under prevalent agricultural and climatic systems, underground pit storage and above-ground silos do not come up to the requirements for maintenance of quality of agricultural commodities. Quality attributes are reflected in bulk densities, production of frass, microbial and entomogenous debris, related mycotoxins and entomotoxins. Biological value, digestibility, and available protein-calorie contents are adversely affected during storage under non-optimum conditions, such as exist in the humid tropics. The chronic and acute toxicities that follow the consumption of deteriorated grains are still not well understood (Howe 1965), but they do possess deleterious constituents and pose health hazards to man and animals. The selective removal of proteins, minerals, fats, and vitamins, the most nutritious components of grains, by insect pests and fungi is on record.

Cereals and millets carry only marginal amounts of proteins, fasts, crude fibres, minerals, and vitamins. However, they are rich in total carbohydrates and in calorie contents. The depletion of specific tissues of the seeds such as the germ, bran, and endosperm have their individual influence on nutritional availability and calorie supply. In tropical and subtropical ecosystems, maintenance of these marginal food and nutritional supplies assumes great importance in view of the profiles of food intake and the delicate balance between nutritional requirements and supplies. These aspects have been very neglected fields of study.

CHOICE OF DISINFESTATION SYSTEMS

In the approach to the maintenance of quality, particularly of the edible portions of the cereal food grain, millets, fruits, and vegetables under tropical and subtropical ecosystems, pre-harvest sanitation prophylaxis and curative measures are not only essential but determine the interface contributions between pre-harvest and post-harvest systems. Pre-harvest prophylaxis with selectively deterrent pesticides for annihilating fungi, insect pests, and mites is absolutely essential (Majumder et al. 1961). Ripe grain panicles are attacked by rodents in the field prior to harvest. They not only consume them but carry away an excess for storage and consumption during the lean season. In order to maintain the theoretical quantum forecast of harvest, rodent and bird repellents are also required. Tropical environments provide continuous sources of infestations by storage pests in the fields prior to harvesting. The proximity of farm houses to agricultural land means a high pest population index because of rodent burrow cultures in fields. Selective compounds with little or no effect on domestic animals and man are available. The technologies and logistics of applications for such formulations are required to ensure planned and integrated systems of pre-harvest and post-harvest agricultural and food productions. These measures can definitely reduce the constraints that now exist for achieving the maximum productivity from land and post-harvest conservation.

Harvested grains must be treated with technologies that are appropriate to the agricultural practices in tropical and subtropical areas. Marginal farmers, subsistence producers, and organized cultivators require technologies for post-harvest drying, threshing, packaging, and storage systems that are in consonance with their economic status and level of skills. Harvesting of panicles for cereals and of haulms for groundnut and tuber crops without mechanical injury is one essential requirement to prevent infection and cross-infestation during post-harvest handling. The development of new technologies and the designing of harvesting tools should, therefore, take into account the physiological and morphological features of these crops.

Drying in cribs using solar energy, oven heat or fire appliances has been the potential contrivance available to average farmers. The good design of such cribs will go a long way towards preventing losses at the threshold of agricultural production. Bringing the moisture to a level that will not encourage storage fungi and mites can surely solve the problem of mycotoxin production in food grains and other crops. Gaseous pasteurization with ammonia, ethyl formate fumigations, and treatment with organic acids have very promising effects on the preservation of the quality of wet harvested commodities (Majumder 1972; Majumder and Venugopal 1973). The effect of ammonia on aflatoxin is fairly well established and disinfection techniques that do not pose any residue or contamination problems can be utilized easily even under field conditions.

SOIL QUALITY AND ENVIRONMENT, AND THEIR INFLUENCE ON POST-HARVEST QUALITY OF FOOD GRAINS AND OTHER AGRICULTURAL PRODUCTS

Apart from natural edaphic factors that influence the agronomical characteristics of crops, the soil environment, which contains pesticides, heavy metals and mycotoxins, seems to have a profound influence on the uptake, translocation and concentration of xenobiotics and other foreign chemicals in the edible tissues of plants. The tissues that are not utilized for human consumption are utilized as fodder and feed for animals, and often the residues are consumed by wild animals and birds. In addition to this inherent translocation, along with solutes that concentrate pesticide residues in tissues, edible by either people or animals, agricultural practices demand the application of pesticides, growth regulators, weedicides and other chemicals during the growth, development, and production of crops. Residues of pesticides lingering in the soil sometimes have adverse effects on crop yields (Karanth et al. 1981).

Preferential uptakes from the soil by different crops have been studied. Some crops have an inherent tendency to concentrate chemicals even beyond the permissible limits prescribed under food regulations. Coriander absorbs much more chlorinated hydrocarbon pesticides from the soil than do amaranthus, knolkhol, chillies, cucumbers, and tomatoes. Tomatoes grown in highly polluted soils showed growth retardation resulting from phytotoxicity, yet there was no actual intake of pesticide in the edible tissues. Other crops like the French bean, brinjal, and ragi were not adversely affected by these pesticide residues in the soil. Paddy exhibited reduced growth when certain chlorinated hydrocarbons were present in the rhizosphere region. Under experimental conditions the half-life of test chemicals was calculated to be of the order of twenty days. The possibility of using French bean, paddy, ragi, and brinjal as rotation crops in soils containing chlorinated pesticide has been suggested. Another approach to mitigate the adverse effect of pesticide residues in the soil is based on the use of crude enzymes, prepared from germinating plants. which have the capability to dehydro-halogenate chlorinated pesticides, including benzene hexachloride (BHC).

The above studies have thus indicated that the decontamination of soils containing high levels of chlorinated pesticides, including BHC residues, might be carried out using certain safe techniques and soil amendements, to achieve crop products free of pesticide residues. If the contents of pesticides in soils are monitored, the antagonistic properties of pesticide combinations could be exploited, both for beneficial effects on plant growth, and also to obtain a remission in the pesticide residue uptake from soils. Combinations, such as gardona + HCH, and fenitrothion + HCH, worked in this way in pot cultures (Deo et al. 1980).

Another phenomenon that has been examined relates to the isomerization of some inactive isomers of chlorinated hydrocarbon pesticides to their active forms. Similarly, active forms could be isomerized to non-active isomers in aqueous medium which could be abiotic or biotic. Fruits or cereal panicles when sprayed with pesticides at recommended dosages may not carry residues beyond the permissible limits, but deposits on the leaf surfaces, sheaths, flowers, and other tissues show a several-fold increase in concentration beyond permissible limits due to the surface/weight ratio.

The leaf is broad and outspread and thus receives heavy deposits of spray. The weight of leaves being smaller in proportion to surface area, the concentration could, arithmetically, go beyond permissible limits.

This factor poses health hazards and related problems to grazing animals. The need for decontamination of fruits, vegetables, grains, and other agricultural crops from pesticide residues prior to consumption is becoming more relevant in today's context (Visweswariah et al. 1977). The water runoff and effluent in a farm environment contain pesticide residues, either in emulsified form, suspended. or in soluble form. Fish culture is posing difficult problems due to the big-magnification phenomenon and high toxicities of pesticides towards them. Sources of water that are contaminated with farm runoff, need purification prior to utilization for drinking purposes, fish culture, algal production, and even for irrigation purposes. The implications of pesticide residues in harvested crops, in their nutritional and toxicological aspects, need deeper studies.

DISINFESTATION IN POST-HARVEST ECOSYSTEMS

In the tropics, post-harvest disinfestation should include two basic features: (a) freedom from infestation from within the seed or grain materials; and (b) effectiveness of the treatments to prevent cross-infestation from outside sources. In order to fit into the interfaces and interphases of agricultural and post-harvest systems, integrated application of the following techniques and products will be required to prevent big-deterioration at all stages of production, processing, handling, storage, and distribution (Majumder 1970). At the same time, it would minimize the risks of pesticide contamination of food and pollution of the environment.

I. Rural Sectors

a. Pre-harvest prophylaxis by spraying of
   i. malathion formulations on grains or pods or synthetic pyrethroids or primiphos methyl prior to maturity and harvesting of grains;
   ii. bacterial insecticides on paddy and vegetables.
b. Field drying and processing
   i. field drier, crib drier, and algate drier;
   ii. maize hand-sheller;
   iii. groundout and paddy drying of haulms and stalks, respectively;
   iv. control of fungi on wet harvested grains.
c. Rodent control in the field with liquid fumigant emulsions and protecting the crop with high-viscosity rat repellent formulations;
d. Insect-proofing and rodent-proofing of structures;
e. Insect-proofing of gunny bags;
f. Rodent control techniques for dwellings; repellent spray, baiting with optical attractants, trap-cum-baiting stations, fumigation of burrows with fumigant emulsion or liquid fumigants. and detoxification of spent poison baits;
g. Spot fumigation with liquid fumigant and fumigant ampoules, fumigant tabloids, fumigant cream;
h. Incorporation of activated kaolin in raw pulses and other legumes' seeds for limiting intergranular space;
i. Addition of nutritional grain protectants based on tricalcium phosphate, vitamins, and glucose on rice, wheat, protective foods, processed cereal foods, breakfast foods, etc.;

j. Introduction of multi-purpose prefabricated unitized gasketed panel structures suitable for storage and gaseous disinfestation;

k. Institution of rural disinfestation services for curative and prophylactic treatments in the preharvest and post-harvest phases of crop production/utilization systems.

The above techniques, which have already been standardized and tested in the tropics, could cover up to 70 per cent of all food-grain production in a county like India. Such pre-harvest and post-harvest integration of disinfestation techniques in the rural sector has the possibility of augmenting food and nutrition supplies to a significantly larger extent than any other effort from existing production.

II. Urban and Household Sectors

a. Warehouses

i. Durofume process that includes a tropicalized fumigant for disinfestation and prevention of cross-infestation. with high-viscosity insecticidal treatment of stacks;

ii. ballooning technique for storage of moisture-sensitive commodities that prevents ingress of moisture from humid atmospheres;

iii. rodentproof design of storage structures.

b. Food-processing Factories and Mill Sanitation

i. Heat disinfestation for processed dry products and Entolation of milled powdered products;

ii. serial fumigation for inpackage disinfestation and inpackage injection fumigation;

iii. nutritional grain protectants for inducing immunity against insect attack by selective metabolic aberration of insects;

iv. multiple fumigation process for mixed commodities in large warehouses;

v. insect and rodent proofing of cartons and packages for dry foods and macaroni;

vi. dehydrobin to mitigate moisture condensation in metal bins during storage of grains;

vii. laminated pouches/cartons;

viii. silo-cum-elevators under large mill and bulk handling conditions;

ix. air curtains in warehouses and processing plants;

x. warehouse rodent proof designs;

xi. entomological sanitation of surroundings.

c. Household

i. Decontamination of fruits, vegetables and grains from pesticide residues;

ii. household spray emulsion and sprayer, repellent spray for domestic pest control;

iii. spot fumigant tablets, paste, and powder for disinfestation of rations in domestic store;

iv. calcium phosphate-based grain protectants for long-term storage of food grains, and processed and formulated products;

v. pyrethroid formulations for store. kitchen and toilets;

vi. anti-termite pre-construction and post-construction treatments with newer products;

vii. flyproofing/mosquito proofing using mechanical or chemical methods;

viii. vector control methods for controlling flies, mosquitoes. etc. with bacterial pesticides.

Special mention should be made of the recent development of non-toxic pesticides, which are either selective deterrents or toxic to insects. These are based on specificities in the physiology, morphology. biochemistry, and immunology of insects, which are, luckily, different from higher animals. Whenever these highly specific pesticides could not be effectively used for the control of insects, moulds, mites, and rodents, other techniques which have ecological compatibility have been selected, and find a place in the above list.

Most of the products and technologies listed do not pose environmental residue and other pollution problems. The basic principle of integrating pre-harvest phases with the post-harvest operations can serve as a prophylactic system. Accordingly, a preventive approach to the problems of big-deterioration caused by fungi, insect, mould, and mites has been the general theme of the present paper. Nutritional augmentation has been inherent in the approach, since grains are protected not only from adverse changes brought about by tropical humid environmental conditions, but from pests which reduce nutrients. The problems of mycotoxins, entomotoxins, calorie depletion, nutritional loss, and related factors have been taken into account in selecting both the particular selective compound and the method of conservation for use at various points in the pre- and post-harvest continuum.

THE USE OF SOLAR ENERGY IN POST-HARVEST TECHNOLOGY

The present crisis of fossil fuel shortages for producing thermal energy is creating increasing emphasis on the use of solar energy. Solar energy is freely available and is abundant in India for most of the year It is non-polluting and inexhaustible. This paper describes different solar energy systems developed in India with their technical and economical feasibility criteria for the post-harvest technology of cereal crops. These include drying, storage, processing, and household cooking in order to save costly fuels now used for these operations.

Post-harvest technology is a multidisciplinary field and includes various treatments and operations carried out on harvested crops for the purpose of preservation or enhancement of quality for marketing and consumption. In these operations, a considerable amount of produce is lost that needs to be minimized. The factors responsible for these losses include inefficient harvesting, threshing, handling, drying, processing, and cooking.

Solar equipment costs are generally high, relative to fossil fuel equipment costs, but operating costs are much lower. The variability of solar energy also limits its usefulness. However, as fossil fuel supplies dwindle and costs continue to rise, solar energy should become economically feasible for many applications in agriculture. With the introduction of high-yielding varieties, fertilizer and better land and water management practices, crop yields have increased substantially (from 52 mill. t. in 1951/52 to 130 mill. t. in 1979/80) in India. However, the benefits of increased crop production are lost if management is lacking to handle this increased production with minimum grain losses during the post-harvest period.

Post-harvest technology is a multidisciplinary field and includes various treatments and operations carried out on harvested crops for the purpose of preservation or enhancement of quality for marketing and consumption. The major operations are threshing, drying, storage, processing, packaging, transport, and cooking. In these operations, a considerable amount of produce (about 10 to 15 per cent for cereal grains and 30 to 40 per cent for pulses) is lost, which needs minimization (Ojha 1976). The factors responsible for the above losses of food grains include inefficient harvesting and threshing, handling, drying methods, poor processing techniques, and inadequate cooking methods.

Solar energy can be used in many of the above post-harvest operations to save costly fossil fuel now in use. Some of these are crop conditioning, hot water production, and steam generation for crop processing and household cooking. Solar-powered cold storage is also technically feasible (Pandya et. al. 1980).

Solar energy is free, non-polluting, and inexhaustible. Based on the data available with the Indian Meteorological Department, the average solar radiation for India is about 420 cal/cm2-day ranging from 323 cal/cm2-day at Gulmarg to 511 cal/cm2-day at Jodhpur. The total energy received in India is estimated to be 60 X 10(16) kwh/year (Bhide 1975).

Although solar energy is free, the equipment required to collect and use it is not. Solar energy utilization devices and processes in general follow one of three pathways; (a) utilizing heat from the sun (thermal devices), (b) converting solar energy directly into electrical energy (photovoltaic), and (c) using photosynthetic and biological processes (biological). Major operations in post-harvest technology use heat from the sun at low temperatures (40°C-90°C). These devices have been developed to a stage at which they can be used on a large scale immediately.

Most of the devices depend upon collection of heat by either a flat plate collector or through concentrating systems and utilize thermal energy to heat a working substance (air, water or some other fluid) to do a desired function. The majority of these thermal devices use flat plate collectors that are simple and stationary. Concentrating systems, although capable of giving higher temperatures, are costly and complicated as they have to track the sun. A well-designed flat plate collector system is capable of attaining temperatures of 80-90°C at 70 per cent collection efficiency.

The development of suitable solar energy systems is a potential means of utilizing solar energy most effectively if based on least-cost criteria designs. Solar equipment costs, relative to fossil fuel equipment costs, have slowed solar development. But as fossil fuel supplies dwindle and fuel costs continue to rise, solar energy collection should be economically feasible for many applications in agriculture. Solar energy can reduce our dependence on fossil fuel, but at the present level of technology its availability is too variable and too limited to completely replace fossil fuel.

This paper describes different solar energy systems developed in India with their technical and economical feasibility criteria for the post-harvest technology of cereal crops. These include drying, storage, crop conditioning, processing, and household cooking.

DRYING

All the cereal crops give higher yields (10-15 per cent) if harvested at optimum moisture content, which is slightly higher than the usual harvest moisture content (Ojha 1976). This reduces shattering losses in the field and transport losses from the field to the threshing yard. It also enables the farmers to plant a second crop after the rainy season crop with the available moisture in the soil (ICRISAT 1981). This early harvest needs adequate drying facilities to reduce grain losses as a result of rotting, discoloration, sprouting of grains, pilferage, bird attack, etc. The most common method of drying cereal crops is to spread the material on the ground or on a raised platform (sometimes on a metal road) in thin layers to expose it to the sun's radiation. Although this method has the advantages of low cost, it suffers from various drawbacks. One has to depend entirely upon the weather conditions so that in rainy or damp weather, drying is unduly delayed causing spoilage of the material. Even in fair weather, drying is non-uniform, because the top layer dries faster than the bottom one. Considerable losses accrue to the produce due to birds and spillage, and contamination occurs with dirt, dust, and surrounding foreign materials.

The drying of cereal grains on small farms can be done either by natural drying or convective drying using heated or unheated air to remove the moisture from the produce. The use of a natural ventilation system is practical for maize cobs harvested at about 20 to 25 per cent moisture content and stored temporarily in cribs (Boshoff 1979). Drying with unheated air is done under favourable weather conditions using heat from the atmospheric air.

Mechanical drying with heated air is of quite recent origin in India and is used on a very limited scale because of the relatively high initial cost (Ramarao 1974). The system consists of an air-heating system (electrical heater, furnace or burner), a fan, and the drying chamber with appropriate airducts. This method offers the advantage of quick, timely, and uniform drying of grains. A mechanical dryer occupies much less space and is suitable to match large-capacity processing plants (Ramarao 1974).

A solar dryer uses a solar energy collector for heating the air. The drying system has separate areas for collection of solar energy and for drying products. The heat is transferred from the air heater to the drying chamber either by convection or forced circulation of air. A flat plate solar collector can be used to dry agricultural produce efficiently and economically during sunny as well as overcast periods (Singhal 1979).

Several types of solar dryers with natural convection as well as with forced convection have been developed and reported in the literature. A solar dryer with natural convection is slower in performance than the forced convection type and thus has more limited capacity (Singhal and Gupta 1975). However, it is much cheaper to operate.

A 0.5-ton low-cost solar dryer has been developed at the Asian Institute of Technology, Bangkok (Exell and Kornsakoo 1979), which is capable of drying paddy in batches, every few days. A similar dryer using metallic construction has been developed at the Energy Center of the Indian Institute of Technology, Delhi. The major drawback of these devices has been that the entire harvest cannot be dried at once and wet paddy waiting to be dried may spoil.

Small portable batch driers have also been developed in India using forced convection. A trolley-type batch dryer has been designed at Punjab Agricultural University, Ludhiana (Pandya et al. 1980).

A 1-ton-per-day-capacity solar dryer has been designed, fabricated, and installed at the Food Corporation of India Modern Rice Mill at Manlur, Chidambaram (Tamilnadu), to dry parboiled paddy. The collector for air heating on the roof is approximately 58 m^2 (8.9 m x 6.6 m). The air flow rate is about 50 m^3/min with an average recorded temperature of 15°C above ambient. Hot air is delivered by an air blower into the drying unit. A 5-ton per day capacity drier based on the above design was installed at the Central Farm Corporation, Ludhiana, Punjab, to dry paddy harvested during the rainy season.

At ICRISAT a solar crop dryer using a single glass-covered flat plate solar air heater was designed during 1977/78 in co-operation with the Mechanical Engineering Department, Punjab Agricultural University, Ludhiana, Punjab. The dryer was designed to dry one ton of grain from a moisture content of 24 to 14 per cent in one day with an overall drying efficiency of 30 to 40 per cent for a solar radiation intensity of 500 cal/cm2-day. The dryer consists of a solar air heater of 50 m2 collector area (10 m x 5 m), a 2.2 kw electrical motor-operated blower, a 1-ton capacity drying bin and ducts to guide the air from the heater to the drying bin. The collector air-heating passage was designed for 1 kg of air per ^{2} per min. The entire solar air heater was mounted facing south and inclined at an angle of about 22° to the horizontal. Due to mechanical problems, only limited tests could be performed on this dryer from November 1978 to June 1979.

The solar dryer was evaluated for its performance on clear days in November and December 1978. Sorghum grain at about 20 per cent moisture content was dried in the month of December 1978. Also sorghum and chick-pea (1 ton each) at moisture contents of 29 and 20 per cent respectively were dried in the months of February and March 1979 to a moisture content of 12 per cent.

A maximum rise in temperature of 21°C above ambient was attained in the month of February; other months gave a rise of 18 to 20°C above ambient temperature at the designed air flow rate between 12 noon and 3 p.m. on days of bright sunshine. The results of sorghum and chick-pea drying. Drying of Sorghum Grain Using Solar Energy. Date = 1-2 March 1979; variety = sorghum CSH-6; ambient temperature = 28 + 2° C; hot-air temperature = 40 ± 5° C, overnight tempering = ·) and 6. The chick-pea was dried from 29 to 12 per cent moisture content during a total drying period of 10 hours (two days) with overnight tempering. The solar air heater efficiencies varied from 51 to 62 per cent depending upon the time of year. The highest efficiency was observed-during December to March, although the outlet air temperature was low during these months compared to hot months like May and June with lowest thermal efficiencies.

The economics of solar drying, compared with the most common methods of drying such as sun drying and mechanical drying using different air heating systems, is presented in table 1. The systems are compared for drying 1 ton of grain from 24 to 12 per cent moisture content in a day. in situations where the systems are used for either 30 days or 150 days in a year on a farm and an agro-industry respectively.

It is clear that sun drying has a low cost of operation as well as low capital investment and hence it is popular, but it does suffer from such inherent drawbacks as spillage of grain from unexpected rains, contamination with dirt and dust, etc. The mechanical dryer using solar energy-heated air comes next with regards to initial investment as well as cost of operation.

TABLE. Economics of Grain Drying Using Different Methods

Method	Initial cost (Rs)	Cost of operation for: 30 days (Rs/ton)	150 days
Open floor sun drying in one day	9,000	48	24
Mechanical drying using artifically heated air	45,000		
Direct oil fired burner		275	76
Coal fired furnace with air heat exchanger		269	78
Husk fired furnace with air heat exchanger		257	54
Husk fired furnace with steam heat exchanger		256	53
Mechanical drying using solar energy	30,000	218	42

Rice is the staple food in India. More than 50 per cent of the paddy produced is parboiled. The basic requirements of parboiling are soaking, steaming, and drying of paddy (Ali Nawab 1974).

A modern parboiling process developed at the Indian Institute of Technology, Kharagpur. utilizes hot water for soaking at 70 to 80° C for 3.5 hours, thus eliminating the steaming process, after which the soaked paddy is dried. Parboiling is done for more than 200 days in a year. It is estimated that with the use of solar energy. a saving of Rs 30,000 (US$3,300) per year can be made for a plant of 1-ton-per-day capacity (Singhal 1979).

A pilot plant of about 100 kg/day capacity for parboiling and drying was designed and tested to determine the feasibility of solar energy utilization (Singhal and Gupta 1974). The parboiling unit consists of a parabolic cylindrical concentrator 3.6 m long and 1.86 m wide, with a focal length of 0.46 m and four well-insulated brick masonary parboiling tanks.

The water heated to the required temperature by the concentrator was directed to the parboiling tanks filled with raw paddy. The mixture was allowed to soak for 3 to 4 hours.

The drying system was designed so that the roof of a farmer's house or small rice mill shed can be converted into an air heater. A flat plate air heater of 10.8 m^2 surface area was used to heat the atmospheric air. The soaked parboiled paddy was dried at an air temperature of 60 ± 5° C with an air flow rate of 15.25 m^3/min.

TABLE. Performance of the Paddy Parboiling System[a]

Date	Time required to parboil[b]			Parboiling system efficiency	Concentrator efficiency	Parboiling efficiency
	From	To				
	(fur)	(fur)	(fur)	(%)	(%)	(%)
IR-8						
23 April	10.30	14.00	3.30	35.5	41.4	85.6
	13.00	17.00	4.00	32.1	41.4	77.7
25 April	10.00	13.15	3.15	37.2	41.4	89.9
Jaya						
25 April	13.00	17.00	4.00	32.0	41.4	77.3
	13.00	16.30	3.30	35.6	41.4	86.1

a. Water temperature: 75 + 2°C, number of tanks: 4 (2 at a time) Insulation thickness: 8 cm; capacity of each tank: 8 kg. Type of insulation: rice husk; size of concentrator: 3.6 m x 1.8 m. Reflection material: aluminium sheet; time of investigation: April 1976.

b. Excluding the time required to heat the water by the concentrator.

Source: Singhal 1979.

TABLE. Drying Performance of Solar Energy Operated Drying Unit

Time of day (h)	Solar intensity (cal/cm² min)	Ambient temperature (°C)	Drying air temperature (°C)	Moisture content (%)	Rate of drying (%)	Drying efficiency (%)
21 May 1976						
12.00	1.40	39.5	56.5	54.36	-	-
13.00	1.34	41.0	59.0	43.38	11.00	97.50
14.00	1.28	42.0	60.5	33.25	9.60	82.60
15.00	1.16	43.0	600	27.40	640	64.40
16.00	0.98	42.0	59.0	22.30	5.10	52.40
Overnight tempering, 22 May 1976						
9.00	1.18	35.5	55.0	23.45	-	-
10.00	1.24	36.5	56.0	18.15	5.30	47.60
11.00	1.25	38.0	58.0	16.50	1.65	14.40
Average efficiency (%)				67.40		
Average air-heater efficiency (%)				65.66		
Overall system efficiency (%)				44.25		

Source: Singhal 1979.

The parabolic cylindrical concentrator was able to heat the water to between 70 to 80° C during the collection period, giving a rise of 50 to 55° C above ambient. The efficiency of heat collection was about 55 per cent.

The average drying rate was 5 per cent moisture loss per hour for a 100 kg grain sample, and the total yield and head yield were 75 per cent and 69 per cent for JR-8 variety paddy. The total drying time was 8 3/4 hours. An average drying efficiency of 67 per cent was attained for the system.

The mechanical strength (crushing hardness), milling index (brokers percentage), optimum cooking time and swelling index for cooked rice of the parboiled rice thus produced were intermediate between raw and hot soaked rice.

TABLE. Quality Data of Rice Obtained by Different Parboiling Methods

Method of parboiling	Crushing hardness	Brokens	Optimal cooking time	Swelling index
	(kg)	(%)	(min)	
IR-8				
Raw	6.10	22.50	25 00	3.71
Solar energy parboiled	6.60	9.80	27 00	3.75
Hot water soaked	7.40	7.52	31 00	3.81
Steam parboiled	8.60	7 00	41 00	3.85
Jaya				
Raw	5.90	34.25	22.00	3.51
Solar energy parboiled	6.40	10.05	23 00	3.70
Hot water soaked	7.20	10.05	23 00	3.70
Steam parboiled	8.20	7.20	45 00	3.85

Source: Singhal 1979.

A proposed layout of a parboiling plant for a mini-rice mill of 2 ton/day capacity, based on the above investigations, is shown in figure 9.

1. Aluminium foil reflector (2 m x 4 m),
2. receiver pipe - 8 cm dia. (black-coated aluminium).
3. 25 mm dia. wheel valve,
4. 25 mm G.I. pipeline well insulated,
5. parboiling tank,
6. thermostat valve,
7. water tank (Singhal 1979))

(A) *Initial investment and area required*

Parboiling

Conventional method Rs 35,000

Solar energy operated Rs 32,000

The effective reflecting surface area for solar energy operated system is 75 m^2 per ton of paddy.

Mechanical drying	
Mechanical dryer	Rs 40,000
Solar energy system	Rs 32,000

Collector area of 100 m^2/ton of paddy at 14 per cent moisture content is required assuming the drying unit efficiency as 40 per cent.

(B) ***Cost of operation***	
Parboiling cost	Rs/ton
Conventional method of parboiling	34
With coal and fuel	34
With husk as fuel	24
Solar energy	19
Mechanical drying (using artificially heated air)	
Steam heat exchanger with coal fired boiler	31
Steam heat exchanger with husk fired boiler	18
Air heat exchanger with husk fired furnace	19
Direct oil fired burner	25
Solar energy operated mechanical drying	14

a. The cost estimates were made in 1979. Source: Singhal 1979.

Storage

Most storage losses of cereal and pulses are caused by insect damage, mould and heating due to excess moisture. Often loss occurs because of moisture accumulation in grain, even though it may be dry when in storage (Shrivastava 1975). It is important that the product is properly stored after it is dried. Stored grain is usually given a chemical treatment to reduce insect infestation for effective storage. Insect infestation can also be prevented by exposing grain to heat treatment or placing it in cold storage. A solar heat treatment machine for destroying initial infestation in maize and other grains has been developed at the University of Udaipur, Rajasthan (Pandya et al. 1980). Technically it is feasible to have a solar-powered cold storage plant to maintain a temperature of 4 to 5° C. but this system has a high initial cost. A 0.5-ton capacity solar powered cold storage unit has been developed at the Indian Institute of Technology, Bombay (Sukhatma et al. 1979).

HOUSEHOLD COOKING

About 150 million tons of firewood are burnt yearly for cooking in the rural areas of India by the poor sections of society (Garg 1978). The ovens and chulas (fireplaces) used are quite crude and inefficient. Due to scarcity of firewood, the rural population is switching to kerosene as the domestic cooking fuel. This tendency needs to be curbed to conserve foreign exchange reserves, which is why solar cookers or ovens are being developed in this country.

India is a pioneer in the development of solar-cookers. The first cooker was produced in India in 1950 and became commercially available in the 1960s but it did not become popular at that time due to the cheap availability of cooking fuels. However, demand is being felt now. The cookers are becoming popular in Gujarat and Maharashtra states. Two models of solar cookers have been developed. The conical shaped sun-basket, made from fibre glass-reinforced plastic (FRP), serves as solar cooker and baking oven by concentrating the sun's rays at a point where a temperature of 110° C can be achieved within a period of 5 to 10 minutes on clear sunny days. The initial cost of this basket is about Rs 600. The reflector needs to be rotated to track the sun.

The hot box type solar oven works as an air-tight box with double glass covers, in which one or more cooking containers can be placed. It requires more time than the sun basket to cook food but it is able to maintain the temperature for a longer duration, making it useful during intermittent sunshine. The initial cost of this oven is about Rs 500. The major problems being experienced in the adoption of these cooking devices are:

1. It is difficult to produce a solar cooker at competitive prices for indoor kitchen use. No housewife prefers to cook in the sun.
2. About 70 per cent of the rural population cook their meals in the early morning before going to work and in the night after returning home. which is not possible by

the use of solar cookers. It will be necessary, therefore, either to change cooking habits or to provide low-cost heat storage.

3. Solar cookers and ovens require more time to cook meals than do conventional methods, and sometimes it becomes difficult to cook even at noon when radiation is very low. This variation is not acceptable to most families.

CONCLUSIONS

The use of solar energy is technically feasible in major post-harvest operations affecting cereal crops. Although the total solar energy available in India is immense, the cost of capturing this in usable form is not small. At the present level of technology, the capital cost is high. There is still scope for modifying solar devices to make them acceptable to the farmers and small-scale agro-industries.

DISCUSSION

Choices of Technology

Many food mixes for children have been described. What are the factors that might dictate their choice? Raw material availability is, of course, the most important of these. Granting this, for home use the technique of puffing, which is a high-temperature (250° C), short-time (one-minute) process, has many attractive features. It is a quick, dry method; moisture is reduced to 2 to 3 per cent, and the traditional use of sand could with advantage be replaced with salt, which does not adhere to the very dry product. Germination is rather tedious for household use, and is better fitted for factory-scale utilization since all the steps of wetting, germination, drying, roasting, and powdering are well adapted to mechanization. For community adoption, all the methods described have potential.

Parboiling is not a universally applicable procedure in milling grains. The main object of parboiling is to harden the grain and thus reduce milling losses. Since wheat and sorghum are almost sure to be ground later into powder for home use, parboiling has less relevance for these grains. Some sorghum varieties are soft, and respond well to parboiling for subsequent preparation of semolina in good yield. Similarly, bulgur wheat represents a successful parboiling variant. Since parboiling achieves case-hardening, it is also an excellent method of prophylaxis prior to storage.

Solar drying is making some commercial headway, and a few devices are being marketed in Maharashtra and Gujarat. For domestic cooking, work was needed to build cookers that could be used in the kitchen itself and not in the open as at present. In the paper, energy costs from coal had been compared to those from furnace oil, but a comparison between solar and gobar gas energies would be relevant.

Technologies for grain storage and drying also offered a wide range of options. The "Save Grain" campaign of the Indian Government had 17 offices and 6 research stations and was operated by all state governments with the aim of educating and motivating farmers through demonstration, training, and publicity. Top priority was given to fabrication of appropriate storage structures and timely supply of pesticides. Funds were furnished to state governments to fabricate bins for the use of farmers through deferred payments or subsidies. Some 300,000 metal bins had so far been sold, besides Rs 1 million worth of ethylene dibromide compounds a year for fumigation. Such inputs were essential in the early stages to move farmers towards adoption.

One could hardly expect definitive figures for grain losses in storage, and there is bound to be a wide range. It is a stupendous task to dry 130 million tonnes of food grains in a matter of a couple of weeks, and the techniques devised should match the quantity of produce. It is common to find insufficiently dried paddy being marketed in Punjab and Haryana with a moisture content as high as 18 per cent, which was disastrous for the consumer.

Traditional home technologies for grain storage frequently failed to recognize losses in quality and quantity, or even potential danger. The use of castor oil as a grain protectant may simply depend on the slipping away of eggs, and other infestations go unrecognized. Mercury pellets have an ovicidal action, but cannot inhibit the growth of larvae and pupae within the grain; if used early enough, they have a value. Mercury was not harmful in itself, but is converted in the body to methyl mercury, which is highly toxic.

The Adoption of Technology

Why has all the excellent work on child feeding not taken root in the system? It is not easy to offer a clear answer. People must be motivated to adopt even something that is meant for their own good. Anything free or subsidized or charitable will not last long. It is the demonstration effort that must be subsidized, and the business of sponsorship is to bring technology to the attention of people, who can be expected to adopt it if the innovation is a commercial proposition, or if it fulfils a real need. Commercial foods cost the buyer about twice what they actually cost to produce, and the types of strategy used to sell them are not likely to be suitable for really low-cost foods. Here, decentralized local production for use over, say, a 40-kilometre radius appears best suited, with minimal packaging, overheads, and commissions. It is true that Indian buying habits are changing, and one cannot predict what will be a commercial success and what will not. Where purchasing power is almost non-existent, as in a village, one approach to making a processed food is to have local women produce and market the product at some small monetary benefit to themselves. Organizations like Mahila Mandals have had some evident success in this direction. But, overall, a commercial approach, however marginal, to adaptive or innovative technologies is inescapable.

VIEWING PROBLEMS

Have problems really been approached with a view to their solution? Where, for example, does solar energy best fit in, and what devices will then meet the ends identified? The interphases surrounding the problem must never be lost sight of. Sometimes a pre-harvest prophylaxis is a better answer than a later intervention. Either the science behind traditional technologies, or even a new technology, must be geared to its greatest need. Weaning foods are probably best directed to the family, to the mother. If a mother simply does not have time to take her child to the immunization centre, the latter must be brought to her.

Perhaps if we examine more closely why people do what they do. some of the many technologies available at the post-harvest stage would have a better chance of passing into the fabric of the nation.

INTERACTION AT THE CONSUMERS' STAGE

To meet the increasing demands of mankind, it is essential not only to grow more food but also to conserve and utilize fully what is grown. Naturally occurring toxicants are present in some foods, and ways and means of minimizing their hazard have to be found to make them suitable for consumption. The approaches toward these include genetic means of selecting varieties that will either contain or support the least amount of toxicants and evolving technological procedures for detoxifying them.

Some of the specific examples of natural/y occurring toxicants causing disease outbreaks in man and approaches toward their prevention will be discussed in detail. These include the problems of (a) lathyrism caused by the consumption of the pulse Lathyrus sativus as staple; (b) veno-occlusive disease outbreak attributed to consumption of minor miller Panicum miliare mixed with weed seeds of Crotalaria, (c) aflatoxicosis caused by consumption of maize contaminated with aflatoxins; and (d) enteroergotism resulting from the eating of pearl millet having ergotized grains.

Although genetic and technological means of preventing/or removing these toxins have been suggested in the past. it is only a package of social management practices that will ultimately minimize the health hazard to man. This calls for strategies of interaction not only between agriculture, nutrition, and food science, but also the involvement of management, sociological, economic, and co-operative sectors.

One way of meeting the increasing food needs of mankind is to fully conserve and utilize what is grown. Utilization of certain foods is hampered through the presence in them of undesirable substances often referred to as toxicants. By trial and error people have selected for their consumption foods that generally do not pose a hazard to them. Yet, for various reasons, population groups are sometimes forced to consume foods that do contain toxicants. The reasons for such forced consumption include easy cultivation and availability, ignorace of toxins, and social and economic compulsions. Several approaches to minimize the hazards from these toxic substances have been evolved. These include the genetic approach of selecting and cultivating identified varieties that contain the least amount of toxin, the agronomic approach of allowing minimal toxin formation during cultivation. and the technological approach of either simple household or large-scale removal of toxins.

This paper describes a few specific examples of naturally occurring food toxicants known to be responsible for disease outbreaks in India.

LATHYRISM

In parts of central India and pockets in Bangladesh and Ethiopia, certain segments of the population are still known to suffer from a crippling irreversible paralytic disease popularly referred to as lathyrism. In the endemic areas of the Rewa, Satna, Durg, and Raipur districts of India, a survey carried out by the National Institute of Nutrition in 1980 indicated that there had been no fresh cases during the last seven years, although there were surviving patients of old cases.

The disease lathyrism has been attributed to the consumption of seeds of Lathyrus sativus as a staple. Cultivation of lathyrus in endemic areas has been reduced in recent years and, consequently, the proportion of lathyrus in the diet has also decreased. While this toxic pulse is also grown in parts of Maharashtra, Gujarat, and Andhra Pradesh, it is used by the local population of these areas to prepare various side dishes and not as a staple.

There are several reasons why the lathyrus crop is cultivated: ease of cultivation; drought-resistant character; good taste; high protein content; a price lower than that of other pulses; and tradition of cultivation in certain regions.

Although attempts have been made in the past to ban the sale of the crop, this approach has not met with much success. In fact prohibition of the "sale, possession for sale and exposure for sale of Lathyrus sativus or use of Lathyrus as an ingredient in the preparation of any article intended for sale" has been made in various states of the Indian Union, though not in Madhya Pradesh, Bihar, and West Bengal following certain provisions in the Prevention of Food Adulteration Act. Thus actual cultivation of lathyrus has not been banned and, in the existing agricultural situation, it would be impracticable to implement any such ban. Some varieties low in toxin that were evolved showed early promise, but the raising of these varieties was never successful for the following reasons: Grains of these low-toxic lines were shrivelled in appearance; seed multiplication facility was lacking; adaptive research in local endemic areas was not carried out; and there was no extension network in the affected areas.

Other methods of control suggested were, first, the detoxification of lathyrus through a simple household procedure, namely boiling in water and discarding the water, or the technological procedure of parboiling, such as is followed in making parboiled rice. These approaches have again not met with success for the following reasons: lack of knowledge; increased cost; increased labour; shortage of fuel; lack of time to detoxify; lack of the facility to dry the produce during the rainy season; practice of mixed crop cultivation, and marketing also as such; different taste of treated seeds; and unsuitability of treated seeds for making chapati.

In view of the drawbacks of both the approaches that were first suggested, social management strategies for lathyrism prevention have been put forward. These include the following:

1. A multisectoral approach involving co-ordination between the agriculture, health, rural development, cottage, and co-operative sectors.
2. Cultivation of alternative crops such as wheat, barley, Bengal gram, lentil, linseed mustard, and cotton.
3. Offering a package of modern agricultural inputs such as supply of seeds, fertilizers, pesticides, spraying equipment, credit facilities, and extension education.
4. Providing increased irrigation facilities.

5. Encouraging cottage industries.
6. Increasing the purchasing power of the local population through the organized collection of minor oilseeds.
7. Levying an additional land cess for the cultivation of Lathyrus sativus.

Adoption of these social management approaches might enable the problem of lathyrism to be contained.

VENO-OCCLUSIVE DISEASE

Veno-occlusive disease is of unknown aetiology, characterized by ascites and pain in the epigastrium, which resulted in the death of more than seventy people, and was reported in the Sarguja districts of Madhya Pradesh. The disease was traced to the consumption of the staple Panicum miliare contaminated with weed seeds of Crotalaria manaburn. Seeds of the weed plants. which grow along with the millet, became mixed up with the millet seeds during harvest and thus entered the human chain. On analysis in the laboratory, the weed seeds were found to contain the toxic pyrrolizidine group of alkaloids. Two approaches have been suggested for prevention of the further occurrence of disease outbreaks.

First, the agronomic practice of weeding the offending plants and, second, the post-harvest method of removing the contaminating weed seeds either by winnowing or sieving. During a subsequent survey it was observed that, even though appropriate sieves were supplied to the households in the affected villages, the wire mesh of the sieves had been damaged by field rats and thus were not serving the purpose for which they were intended. This is another example of how an interaction of interdisciplinary aspects will not be effective if not properly supported by sociological measures, which must include an educational component (Krishnamachari et al. 1977).

MYCOTOXICOSES

Two major mycotoxicoses, namely aflatoxic hepatitis and enteroergotism, were reported in parts of India during 1974-1976. Aflatoxic hepatitis claimed more than 100 people, with symptoms of jaundice, rapidly developing ascites and portal hypertension. It was traced to the consumption of maize heavily contaminated with aflatoxins (Krishnamachari et al. 1975). Maize serves as the staple food of the tribal population during certain parts of the year and was found to be improperly harvested and stored. It was noticed that the maize cobs were either:

(a) kept in heaps covered with grass for a week;

(b) stored in heaps inside the house; or

(c) preserved along with the plant in open heaps in the fields.

Early shelling of the grains was never or rarely practised, and sometimes maize on the cobs was stored in situ in open cribs, either in open yards or inside the house, for several months. As a result the maize grains were affected with moulds.

The inhabitants were aware of the spoiled nature of the grain and resorted to one of the following practices:

1. Spoiled maize was separated out and good ones were consumed immediately. Spoiled maize was stored for later use in emergency.
2. Spoiled maize was consumed first with a view to preventing an imminent loss of maize through spoilage, and good maize was consumed later.
3. Spoiled maize was distributed to agricultural labourers in lieu of wages.
4. Spoiled maize was sold either in the village market or to the poorer segments directly.

Clearly, the poorest sections of the population were exposed to the greatest risk through habitually consuming unwholesome maize because of its low cost. The sequence of events that led to the disease outbreak could be summarized as follows: The problem of aflatoxin contamination could be minimized by agricultural action programmes. These could include evolving and cultivating suitable varieties of crops that will support the least amount of fungal contamination; manufacture and distribution of cheap, easily operated machinery to achieve satisfactory drying; shelling and storage; and pursuit of a set of recommended practices that would prevent mould damage of agricultural commodities. The problem of ergot contamination of bajra, which causes another type of mycotoxicosis, namely enteroergotism, has quite a different perspective. The agro-socio-economic structure of the affected groups is very different from that of the population affected by aflatoxic hepatitis. Ergot contamination of bajra has become more widespread following the introduction of high-yielding varieties - traditional bajra varieties were more resistant. The disease enteroergotism, occasionally reported from the rural areas of Maharashtra, Gujarat, and Rajasthan, generally occurred following the harvest of bajra, and also simultaneously in the affected villages. The clinical picture following bajra consumption was characterized by nausea, repeated vomiting, giddiness followed by drowsiness, and a prolonged phase of sleepiness that extended over 24-48 hours. Symptoms usually started within an hour following the meal, and a single meal was enough to cause the disease (Krishnamachari and Bhat 1976).

The occurrence of the disease could be prevented by the cultivation of resistant varieties of bajra, or through suitable agronomic practices, such as changing the sowing season. Once contamination of harvested bajra is noticed, decontamination measures such as winnowing, sieving, and hand-picking could be resorted to. The salt flotation technique of washing ergotized bajra in a 5 to 10 per cent common salt solution has been successfully utilized by many villagers in the affected areas.

KARNAL BUNT WHEAT

Subsequent to the popularization of certain Mexican dwarf wheat varieties in India, the disease Karnal bunt of wheat caused by the fungus Neovossia indica had in recent years assumed epiphytotic proportions in parts of north-western India. The affected grains lacked lustre and were black in colour, especially at the tip and all along the longitudinal furrow on the central side. Severely affected grains retain only the outer pericarp and the aleurone layer, and hence the name bunt. The affected grains emit an odour similar to that of rotten fish because of the presence of trimethylamine, which is known to pose hazards to human and animal health (Bhat et al. 1978). Hence there was considerable anxiety as to whether the use of Karnal bunt wheat was safe. The nutrient composition of Karnal bunt wheat grains was similar to that of normal wheat, except that the thiamine and lysine contents of affected grains were appreciably lower. Short-term toxicity studies in rats fed up to 50 per cent of Karnal bunt wheat over a period of 45 days, and acute toxicity studies in chicks by oral administration of various extracts of such affected wheat, did not indicate any adverse effect (Bhat et al. 1980). Short-term toxicological studies carried out in monkeys fed a diet containing 70 per cent Karnal bunt wheat did not also indicate any adverse effects, as judged by weight gain. food intake, various haematological parameters and liver enzyme levels. Based on these studies, a tentative tolerance limit of 1 per cent Karnal-bunt-affected grains in good grains, which is within the overall limit of 5 per cent damaged grains, has been suggested (Bhat et al. 1981).

CONCLUSION

Although genetic and technological means of preventing or removing toxins from various food grains have been suggested in the past, it is only a package of social management practices that will ultimately minimize the health hazard to man. This calls for strategies of interaction not only between agriculture, nutrition, and food science, but also the involvement of management, sociological, economic, and cooperative sectors of the total system.

8

The Nutritive Value of Foods

The nutritive value of food is modified by processing. The common household methods of food processing are milling, germination, cooking, and fermentation. Milling, in which the coarse fibrous bran or seed coat of the grain is removed, results in significant nutrient losses particularly of B-vitamins and minerals.

The extent of the losses depends upon the degree of milling and the distribution pattern of nutrients in the grain. In milled products, like refined flour, pearled sorghum, and dhal, the big-availability of iron is better than the corresponding unmilled grains which is attributed to the removal of interfering substances such as phytate, tannin, and fibre. Germination or malting increases the vitamin C and folic acid content of food legumes and also degrades the anti-nutrients present in these food grains.

Consequently, availability of iron in germinated grains improves significantly, especially in malted bajra and ragi. Oligos accharides, the components responsible for flatus production present in pulses, are practically completely degraded on germination and germinated pulses are likely to be less flatuspromoting than raw pulses. Cooking in boiling water or by steam pressure is yet another common household practice of food processing.

Apart from making food palatable and safe, cooking inactivates practically all the anti-nutritional factors that are heat labile. Heat stable components like tannin in pulse legumes leach out in the cooking broth. Some B-vitamins like riboflavin and vitamin B-6 are also lost in the cooking of pulses. The growth-promoting properties of the foods, however, are far better than raw foods, and this is attributed not only to the destruction of anti-nutritional factors but also to better utilization of nutrients like proteins and carbohydrates. Dissemination of this knowledge of the nutritional significance of the simple methods of food processing at household level is important.

Nearly all foods, whether derived from plant or animal sources, are rarely fit for direct consumption and require some prior processing. Methods of food processing have been developed over the centuries and are adopted apparently to make the final product more attractive in flavour, appearance, taste, consistency, and so on. Besides these aspects of consumer preference, several of the methods aim at (a) making the food safe and wholesome, and (b) conserving available food for later use by increasing its shelf or storage life.

For a majority of the population, cereals, millets, pulses, and tubers in the diet are the main sources of most essential nutrients. The common household practices of processing these foods include milling, germination or sprouting, malting, fermentation, and cooking. Each of these processes qualitatively and/or quantitatively modifies the nutritive value of the food. Information on this aspect of food processing is important from both the nutritional and public health points of view. A brief review is presented here of the studies carried out at the National Institute of Nutrition on the effect of certain household practices on the nutritive value of foods.

MILLING OF FOOD GRAINS

Use of whole grain flour for the preparation of chapati from wheat, or roti from sorghum and bajra is a common household practice. On the other hand, brown or unmilled rice is rarely consumed. The most frequent form in which food legumes or pulses are consumed is as the decorticated product known as dhal. From a nutritional point of view both these practices, namely using grains with or without milling, and refining, have several advantages and disadvantages. In the process of milling food grains, the main objective is to remove the coarse fibrous bran or the seed coat. However, in all these processes the nutrient-rich parts of the grain, namely the germ and the aleurone layers, are also displaced resulting in a product poorer in nutrient content. The extent of nutrient losses - particularly of vitamins and minerals - during milling of wheat, rice, sorghum, and pulses is very large.

The large variations in the nutrient losses are attributed, firstly, to the uneven distribution of these nutrients in the structural parts of the kernel and, secondly. to the degree of milling. Studies on the trace element composition of wheat flours of different extraction rates have shown that a major proportion of the phosphorus, magnesium, chromium, zinc, and manganese is located in the bran (Christensen et al. 1976; Sankara Rao and Deosthale, 1981) of the wheat kernel. Removal of the bran during milling results, therefore, in significant nutrient losses. Parboiling of the grain is a suggested remedy to minimize the milling losses of thiamin, riboflavin, and niacin in the rice grain (Kik 1957). During the parboiling process, water-soluble nutrients are driven from the outer layers into the inner layers of the grain and thus escape removal during milling. This beneficial effect of parboiling is observed also for chromium, molybdenum, and manganese (Deosthale et al. 1979). Milling losses of these nutrients are therefore 10-20 per cent lower in parboiled grain than in raw rice.

Although the nutrient content of food grains is relatively poor after milling, the big-availability of certain nutrients appears to improve considerably. For example, the bioavailability of iron, as judged by the percentage of it present as an ionizable fraction (Narasinga Rao and Prabhavathi 1978), was found to improve significantly by milling wheat to refined flour and decorticating chick pea to dhal (Prabhavathi and Narasinga Rao 1979). Similarly in sorghum, the ionizable iron fraction increased from 18.6 to 28.7 per cent indicating a 40 per cent rise in the availability of iron (Sankara Rao and Deosthale 1980). Complex polysaccharides of the fibrous bran, polyphenolic compounds such as tannin, and phytate are all known to limit the availability of iron (Hussein and Patwardhan 1959; Disler et al. 1975). Removal of these complex substances by milling partly explains the improved iron availability of milled grains.

TABLE. Soluble and Ionizable Iron Content of Some Unmilled and Milled Cereals and Pulses

Foodstuff	Fe in grain (mg/100g)	Fe at pH 7.5 Soluble (%)	Ionizable (%)
Wheat flour			
Whole	6.1	5.9	4.3
Refined	1.8	13.2	8.2
Sorghum			
Whole	4.2	-	19.6
Pearled	3.7	-	28.7
Chick-pea			
Whole	6.0	12.6	2.7
Dhal	4.9	22.6	14.0

GERMINATION AND MALTING

Soaking in water overnight followed by germination or sprouting of the grain is yet another very common household practice, especially for the processing of pulses.

During germination several enzyme systems become active and bring about profound changes in the nutritive value of pulses (Subramanian et al. 1976; Ismail Noor Mohd et al. 1980). Vitamin C, which is practically absent in dry legume seeds, increases in significant amounts after germination (De and Barai 1949; Prabhavathi and Narasinga Rao 1979). Similarly Babu (1976) has found two to threefold higher values for folic acid in the germinated grains than in the raw grains of the chick-pea and ragi.

On the other hand, it has been reported that anti-nutritional factors such as phytate, trypsin inhibitor, and haemaggiutinins are broken down on germination (Ready et al. 1978; Subbulakshmi et al. 1976).

Phytate, which constitutes over 60 per cent of the total phosphorus in the raw grains of the Bengal gram, drops to a level of 44 per cent in the 48-hour germinated grain with no change in total phosphorus (Prabhavathi and Narasinga Rao 1979). On malting, a process that likewise involves germination, loss of phytate is quite pronounced in both ragi and bajra (Sankara Rao and Deosthale, unpublished).

TABLE. Folate and Ascorbate Content of Germinated Food Grains

		Germination period in hours			
		0	24	48	72
Ascorbate (mg per 100 g)	Chick-pea	0	20	24	30
Total folate (mg per 100 g)	Chick-pea	21	306	344	374
	Ragi	14	16	63	84

TABLE. Effect of Germination, Malting, and Baking of Bread on Phytin-P

	Phytin-P as percentage of Total P	
	Initial	Final
Chick pea 48 hours germinated	62	44
Malted bajra	46	24
Malted ragi	41	33
Leavened bread	52	24

Recent studies have shown that the most commonly consumed pulses, namely chick-pea pigeon-pea, green gram and black gram, contain a significant amount of tannin that is mostly present in the seed coat. In overnight soaking 25-50 per cent of this tannin is lost, probably through leaching. In 24 to 48-hour germination, there is an additional loss of 10 to 25 per cent (Udayasekhara Rao and Deosthale, unpublished).

TABLE. Effect of Soaking, Germination, and Cooking: Percentage Loss of Polyphenols of Four Indian Pulses

		Percentage loss of tannin on:			
	Tannin	Soaking	Germination	Cooking	
	(mg/100 g)	24 hours	24-28 hours	Raw	Germinated
Chick-pea	179 + 21	50	3-8	70	75
Pigeon-pea	996 + 93	50	3-10	60	75
Green gram	612 + 53	25	20-25	70	76
Black gram	861 + 92	25	10-25	70	77

The beneficial effect of these changes in food grains on germination and malting are reflected in the improved big-availability of nutrients. Thus the ionizable iron content in chick-pea and wheat progressively increases on germination up to 72 hours. On removal of the seed coat, the ionizable iron is 14 per cent both in the raw and germinated grains of the chick-pea. The absence of any germination effect may be attributed to factors associated with the seed coats, which probably are progressively eliminated during germination. In ragi and bajra also, germination or malting enhances the iron availability about eight to twelve-fold (Sankara Rao and Deosthale, 1981). In the context of widespread iron-deficiency anemia in the population, and the suggested use of malted grains for the feeding of infants and children, these observations on iron availability of malted bajra and ragi are of considerable significance.

TABLE. Ionizable Iron in Germinated and Malted Food Grains

Food grain		Total iron in grain (mg/100 g)	Ionizable iron at pH 7.5 (%)
Chick-pea	Raw	6.9 (4,9)[a]	2.7 (14.0)
	Germinated 48 hours	6.9	5.6 (14 0)
	Germinated 72 hours	6.9	6.9 (14.0)
Green gram	Raw	5.0	5.7
	Germinated	5.0	4.6
Wheat	Raw	4.6	4.1
	Germinated 48 hours	4.6	67
	Germinated 42 hours	4.6	7.2
Bajra	Raw	7.2	9.0
	Malted	3.7	73.5
Ragi	Raw	3.9	7.4
	Malted	3.4	88.3

a. Seed coat removed.

Germination also modifies the carbohydrate component of the food grains (Subramanian et al. 1976; Lebaneiah and Luh 1981). It is well known that consumption of pulses leads to flatulence. Inclusion of legumes in the diet at levels that provide 20-25 per cent of total calories results in a manifold rise in the amount of gas produced in the intestines (Steggerda and Dimmick 1966). Studies in humans fed different pulses have shown that chick-pea is more gas-forming than other pulses (Narayana Rao et al. 1973). One of the factors associated with flatulence is the high concentration in pulses of certain oligosaccharides of the raffinose family. Because of the absence of suitable digestive enzymes these sugars are not utilized by man (Gitzelmann and Auricchio 1965).

The unabsorbed sugars are acted upon by the micro-flora of the large intestine resulting in gas production (Calloway et al. 1966; Richards et al. 1968). Studies on common pulses like the chick-pea, green gram, black gram, and red gram (Udayasekhara Rao and Belavady 1978) have indicated a continuous fall in oligo-sugar content on germination.

In grains germinated for twenty-four hours the oligo-sugar levels are at 50 per cent of their initial value, and by 48-72 hours at less than 25 to 15 per cent. These observations suggest that sprouted pulses are likely to be less flatus-producing.

COOKING

Food is cooked by heat treatment in several ways, which include boiling in water, pressure steaming, roasting, frying, and baking.

Cooking is important to make the food safe by killing contaminating bacteria, and also to inactivate several heat-labile anti-nutritional factors present in many foods. Among the heat-stable factors in food legumes, tannin appears to leach out into the broth during cooking in water and thus the level in the grain is reduced to 30 per cent of the initial value (Udayasekhara Rao and Deosthale, unpublished).

TABLE. Cooking Losses of Riboflavin and Total B6 Content of Dhals

	Riboflavin		Total B6	
Dhals	**mg per 100 g**	**Cooking loss (%)**	**mg per 100 g**	**Cooking loss (%)**
Chick-pea	0.23	35.0	0.38	15.2
Pigeon-pea	0.25	17.8	0.32	13.2
Green gram	0.24	14.8	0.31	12.9
Black gram	0.25	13.2	0.24	12.8

Cooking also results in the loss of vitamins. Studies reported by Raghunath and Belavady (1979a) have shown that cooking loss of riboflavin varies from 13 to 35 per cent in the four pulses. On the other hand cooking losses of vitamin B6 in these pulses were relatively low (13 per cent) and constant. However the availability of food iron is not altered by the cooking of rice and dhal, or by baking a phulka from whole wheat or refined flour. In baking bread, the rise in ionizable iron content is attributed to the prior fermentation of the ingredients (Prabhavathi and Narasinga Rao 1979).

TABLE. Effect of Cooking on Ionizable Iron

	Percentage of ionizable iron at pH 7.5	
Type of process	**Raw**	**Processed**
Cooking		
Rice	9.1	10.6
Pigeon-pea dhal	16.0	16.0
Baking		
Phulka (whole wheat)	3.5	3.6
Phulka (refined flour)	8.2	9.2
Bread (whole wheat)	4.3	9.7
Bread (refined flour)	8.2	20.7
Fermentation		
Idli	2.5	2.5
Wheat flour dough	6.0	6.9

Cooking of food improves the biological value of its protein. Udayasekhara Rao and Belavady (1979) have shown that the growth performance of rats was better on cooked than on uncooked or raw diets based on different pulse and tuber proteins.

This effect was observed to be particularly pronounced in diets based on the winged bean and its tuber. Animals fed raw winged bean or its tuber failed to grow and showed a significant loss in body weight, and those fed the tubers even died.

TABLE. Growth Performance of Rats Fed Raw and Cooked Diets Based on Different Sources of Proteins

	Raw diet (body weight (g))			Cooked diet (body weight (g))		
	Initial	Week 1	Week 3	Initial	Week 1	Week 3
Pigeon-pea	80	78	80	80	99	120
Soybean	84	79	113	84	92	178
Winged bean	81	63	50	87	112	154
Potato	70	84	90	73	77	91
Winged bean tuber	71	65	Dead	71	78	92

TABLE. Net Protein Utilization (NPU) of Casein Diets containing Starches from Different Sources

Source of starch in the diet	Food intake	Body weight gain	NPU of casein
	9 per day		
Maize			
Raw	7.8	2.1	0.68
Cooked	8.2	1.9	0.72
Pigeon-pea			
Raw	7.1	0.1	0.49
Cooked	8.1	1 .3	0.57
Green gram			
Raw	7.8	1.6	0.66
Cooked	8.4	2 0	0.68
Potato			
Raw	9.0	1.0	0.51
Cooked	8 8	2.0	0.60

Improved growth responses to cooked diets were attributed to the destruction of the anti-nutritional factors present in these foods (Raghunath and Belavady 1979b). The utilization of protein is also influenced by the starches isolated from different sources, as shown by Nageswara Rao and Narasinga Rao (1978). These studies further showed that cooking of the diet significantly improved the net protein utilization of casein when the diets contained legume and potato starches.

CONCLUSION

The practical significance of simple household methods of food processing is obvious to those working in the field of food science and nutrition. There is a great need, of course, to make the consumer aware of the nutritional dimensions of such food processing, which has been discussed in this paper. Education and extension personnel, who are juxtaposed in the interface between the consumer and the food scientist, have a significant role to play.

9

Storage and the Quality of Grain: Village-level Studies

Insect pests, aflatoxins, nutritive value. and cooking quality of rice, sorghum, millets, and legumes stored in 100 rural households of Andhra Pradesh for family consumption over a period of three to twelve months were studied. Regional differences in the level of infestation and associated quality changes were noticed. Uric acid levels were found to be within permissible levels in rice, sorghum, and millets, but higher in legumes. A progressive decrease in starch content (of 3 to 11 per cent) and an increase in reducing sugars /of 30 to 70 per cent) were observed in all grains, but varied depending on the grain. In vitro carbohydrate digestibility increased during storage of bajra and groundnut (20 per cent), but decreased by 1 to 5 per cent in legumes. Protein content of the grains was invariably reduced, though the magnitude varied (2 to 11 per cent) depending on the grain and length of storage. Extensive loss of thiamine and considerable losses of niacin occurred in storage of all the grains. Both desirable and undesirable changes in the cooking quality of rice and legumes were noticed, but no such perceivable changes are found in sorghum and millets.

INTRODUCTION

Food grains after harvest are usually stored till the next harvest season for home consumption. It is well documented that about 80 per cent of the food grains produced in India are stored by rural families for their own consumption. Bulk storage is used for the remaining 20 per cent of the grain, which goes to urban markets. Proper storage of the food grains in rural families then would not only help in bridging the existing marginal food gap, but would also contribute to health and nutrition by conserving grain quality. Despite its importance, information on grain quality changes during storage, especially in home storage, is scarce.

Reported storage losses vary widely between 5 and 50 per cent (Swaminathan 1977). Most of these estimates are based on studies conducted in bulk storage structures. A study conducted by Boxall et al. (1979) on farm-level storage of paddy in coastal Andhra Pradesh is one of the few attempts to assess losses in farm-and home-level storage. Information on storage losses - both quantitative and qualitative - of food grains other than wheat and rice, such as sorghum, millet, and legumes, is extremely limited. And even this limited knowledge is based more on laboratory experiments than on field study.

Reduction of weight losses in bulk storage of grain is directly summable in financial terms, and the cost-benefit ratio is a highly convincing factor for research and development in improved storage structures. In contrast to the problems of commercial storage, those of farm and home storage have received low priority because the damage is insidious and often difficult to quantify.

Grain deterioration in storage is caused mainly through (a) big-deterioration, (b) insects and pests, and (c) moulds and fungi. Bio-deterioration is due to the activity of enzymes present in the seed. The extent of deterioration depends upon the level of enzyme activity, which in turn is determined by moisture and temperature.

Insect pests are a greater problem in regions where the relative humidity is high, but temperature has the greatest influence on insect multiplication. At temperatures of about 32° C, the rate of multiplication is such that a monthly compound increase of fifty times the original number is theoretically possible. Thus fifty insects at harvest could multiply to 312 million after four months. Growth of insect pests and moulds raises both temperature and moisture, and thereby accelerates the activity of the enzymes, which would remain at a low level if conditions of storage were favourable.

Fungal attack in storage generally occurs where drying has been inadequate, or where large numbers of insects are present, thereby causing a temperature rise in the grains, or where the stored crop is exposed to high humidity or actual wetting.

All these circumstances lead to grain deterioration during storage resulting in weight loss.

In addition to weight loss, enhanced big-deterioration results in an enhanced loss of nutrients and contamination with anti-nutritional factors. In populations where most of the dietary needs are derived from cereals and legumes, nutritional impairment in these grains during storage will be of great consequence to the health and nutrition of the people.

Several efforts were made in the recent past to summarize the available information on post-harvest losses (Shulten 1982; Tyler 1982) to identify location, causes, and magnitude, and to deduce appropriate strategies for conservation of food grains.

These reports clearly indicate that qualitative changes in storage of these grains, especially at farm and home levels, are the main grey patches in our knowledge of post-harvest damages occurring in food grains.

In assessing damage, emphasis is frequently on weight loss followed by kernel damage. Other forms of damage, such as reduction in quality and nutritive value, viability of seeds, microbial spoilage, and contamination with substances harmful to health or unacceptable for edible purposes, which could be of greater importance than weight loss, are often ignored or given low priority.

Even when these factors are given importance, lack of approved and standardized methodology for assessing qualitative changes is the main constraint.

Recognizing the need for such information, an attempt has been made by the Department of Foods and Nutrition in Andhra Pradesh Agricultural University to study the qualitative changes in farm-and home-level storage of grain, mainly sorghum, millets, and legumes, in selected villages of Andhra Pradesh.

QUALITATIVE CHANGES IN STORAGE OF CEREALS AND MILLETS

Uric Acid and Aflatoxins

In one study (Pushpamma and Chittemma Rao 1981), samples of sorghum, bajra and ragi were collected periodically from households that stored these grains over a minimum period of nine months. Insect damage was maximum in sorghum, leading to maximum weight losses (2.4 per cent) in this grain compared with bajra and ragi. Moisture content also went up as the level of insect infestation and duration of storage increased. In the three grains maximum increase in uric acid content was observed in sorghum, but the level was still below the specified safe limit (10 mg/100 g).

TABLE. Major Constituent Fractions, Weight Loss, and Increase in Uric Acid Content of the Stored Sorghum and Millets

Foods	Storage period (months)	Undamaged fraction (%)		Damaged fraction (%)		Other fractions (%)	% Weight loss	Increase in unic acid (mg %)
		By wt.	By no.	By wt.	By no.	By wt.		
Jowar	0	94	97	3	3	3	0.2	0.0
(Sorghum	5	91	94	5	6	4	1.5	4.3
vulgare)	9	90	89	9	11	1	2.4	5.4
Bajra	0	98	100	0	0	2	0.1	0.0
(Pennisetum	5	94	98	2	2	4	0.2	3.3
typhoideum)	9	94	95	2	4	4	1.0	3.6
Ragi	0	97	99	0	1	3	0.0	0.0
(Eleusine	5	95	99	0	1	5	0.0	1.4
coracana)	9	93	99	0	1	7	0.1	1.6

In another study (Pushpamma and Uma Reddy 1979) on the changes in the quality of rice and jowar stored for up to one year in three different agro-climatic regions of Andhra Pradesh, insect infestation increased progressively in both grains during storage. Jowar samples were more frequently infested than were rice samples stored under similar conditions. Weight losses ranged between 3.9 and 5.10 per cent at the end of twelve months of storage. Samples from the Andhra coastal region had a higher level of insect infestation and higher uric acid levels. But, except for rice samples stored for one year in the coastal region, all the jowar and rice samples had uric acid contents below the safe level.

Though freshly harvested samples of rice and jowar were free of aflatoxin, after three months of storage, six samples of rice were contaminated with aflatoxin and two were above the safety levels.

After twelve months of storage no increase was observed in either the number of samples bearing toxins, or the level of aflatoxin in infected rice samples. But in jowar, both the number of samples and the levels of aflatoxin increased up to six months of storage.

This could be because of the higher prevalent moisture content during storage of jowar, which increased even more through insect infestation. A greater number of coastal area samples were contaminated with aflatoxin than those from the other two regions.

Carbohydrates

In the same study it was observed that length of storage had no significant effect on the starch content of either rice or jowar. Under normal conditions of storage the starch content of rice was unaffected. The reducing sugar content of rice and jowar increased significantly ($P < .05$) during storage. The increase was greater in the first six months than in the next five months.

In vitro digestibility of the carbohydrates of bajra and groundnuts increased on storage for six months, the mean values being 24.2 and 26.1 per cent respectively (Sathyavathi Devi 1978).

Proteins

Data presented in table 2 show that during nine months storage maximum protein losses were observed (11.0 per cent) in bajra followed by sorghum (10.6 per cent) and ragi (5.3 per cent).

There was a progressive increase in non-protein nitrogen content, which could be due both to the presence of insect excrete and protein degradation during storage (Pushpamma and Chittemma Rao 1981). In another study Pushpamma and Uma Reddy (1979) reported a loss of 0.59 to 0.70 g protein/100 g in rice and of 0.51 to 0.74 g in jowar.

TABLE. Moisture, Protein, and Non-protein Nitrogen Content of Fresh and Stored Sorghum and Millets[a]

Foods	Storage period (months)	Moisture (%)	Protein (g)	Non-protein nitrogen (mg)
Jowar	0	10.4	8.5	326
(Sorghum vulgare)	5	10.4	8.2	240
		(0.0)	(- 3.5)	(+ 1.7)
	9	11.1	7.6	246
		(+ 6.7)	(-10.6)	(+ 4.3)
Bajra	0	9.3	10.0	282
(Pennisetum typhoideum)	5	11.0	9.9	285
		(+ 18.3)	(-1.0)	(+ 1.1)
	9	10.7	8.9	297
		(+ 15.1)	(-11.0)	(+ 5.3)
Ragi	0	10.9	7.6	193
(Eleusine coracana)	5	10.9	7.4	216
		(0.0)	(-2.6)	(+ 12.0)
	9	11.6	7.2	275
		(+ 6.4)	(-5.3)	(+ 42.5)

a. Figures in parentheses represent percentage decrease (-) or increase (+) from the initial sample values

Many investigators have reported a decrease in free amino acids and free amino nitrogen during storage of rice and jowar (Pushpamma and Uma Reddy 1979; Iwasaki and Tani 1967), which appears to be dependent on moisture and storage temperature.

There was a gradual decrease in limiting amino acids in stored paddy and jowar, the decrease being greater in lysine content than in other amino acids even when stored under controlled conditions (Uma Reddy 1981). A varietal difference was also observed (paddy 5-10 per cent, sorghum 25-30 per cent). Thus the grain itself appears to utilize some amino acids for its own metabolic activities, and also to consume them through non-enzymatic Maillard browning. Sorghum is naturally deficient in lysine content, and the drastic loss observed following insect infestation will affect the overall biological quality. Local varieties retained a higher amount of amino acids during storage than did hybrid varieties.

Both ß-amylase and protease activity also decreased during storage for 12 months of both rice and jowar in rural Andhra Pradesh (Pushpamma and Uma Reddy 1979).

The decrease in ß-amylase activity was maximum in the first 3 months of storage. where the loss was nearly 60 per cent. Similar results in rice have been reported by many investigators (Srinivasan 1939; Narayana Rao et al. 1954).

Changes in the Biological Quality of Proteins

Assessment of various food grains through the net protein ratio and digestibility coefficient indicated a gradual decrease during storage, Among the cereals, rice retains its protein quality better than does sorghum.

The length of storage had a significant effect on the net protein ratio and digestibility of all varieties of sorghum. In varieties CHS-1 and CSH-6, where infestation is heavy, the net protein ratio decreased drastically from 1.91 to 0.41 and from 2.00 to 0.33 respectively in 12 months of storage. The decrease was comparatively less in local varieties.

Vitamin Losses

A study conducted in rural areas in Andhra Pradesh (Pushpamma and Chittemma Rao 1981) revealed that B-vitamins of all grains were reduced under storage, but the percentage decrease varied from grain to grain depending upon the level of insect infestation.

Maximum loss in thiamine (43.2 per cent) was observed in Eleusine at the end of nine months of storage, followed by Pennisetum (39.4 per cent). The percentage loss of thiamine was observed to be higher during storage than that of riboflavin and niacin.

The rate of decrease of B-vitamins was observed to be highest between five and nine months of storage after harvest. Factors other than insect infestation also contribute, such as moisture, temperature, and oxygen fluctuations, which lead to thiamine losses (Bayfield and O'Donnell 1945; Williams 1952).

TABLE. B-Vitamin Content of Fresh and Stored Sorghum and Millets[a]

Foods	Storage (months)	Thiamine (mg)	Riboflavin (mg)	Niacin (mg)
Jowar	0	0.32	0.18	2.3
(Sorghum vulgare)	5	0.31	0.16	2.1
		(-3.1)	(-11.1)	(- 8.7)
	9	0.24	0.16	2.0
		(-25.1)	(-11.1)	(- 13.0)
Bajra	0	0.33	0.21	2.4
(Pennisetum typhoideum)	5	0.29	0.21	2.4
		(-12.1)	(+ 0.0)	(+ 0.0)
	9	0.20	0 21	2.0
		(-39.4)	(+ 0.0)	(-16.7)
Ragi	0	0.37	0 19	1.3
(Eleusine coracana)	5	0.33	0.18	1.3
		(-10.8)	(- 5.3)	(+ 0.0)
	9	0.21	0.17	1.1
		(-43.2)	(-10.5)	(-15.4)

a. Figures in parentheses represent percentage decrease (-) or increase (+) from the initial sample values

Cooking Quality of Stored Cereals and Millets

Cooking quality is one of the important aspects of food quality. However nutritious a food material may be, it will not be accepted by the consumer unless it satisfies specific culinary characteristics.

Thus the grains of cooked rice should be non-sticky and well separated, qualities that can be judged by swelling number, water uptake, and amylose content. An improvement in the cooking quality of rice was found after six months of storage, but not after three months (Pushpamma and Uma Reddy 1979). It is interesting to note that while the swelling number increased, the amylose content of rice remained constant. Storage decreased the amylose dispersion, which is responsible for greater absorption and retention of water, and thus the expansion of rice grain. The cooking quality of jowar did not improve under storage as markedly as rice. No relationship was noticed between the swelling number and amylose content of jowar during storage.

QUALITATIVE CHANGES IN STORAGE OF GRAIN LEGUMES

Uric Acid and Aflatoxin

A study of rural families conducted in Andhra Pradesh had revealed that after nine months of storage, the percentage weight loss was 1.5 to 2.1 for pigeon-pea, green gram, and black gram, but only 0.5 per cent in chick-pea (Pushpamma and Chittemma Rao 1981). The greatest proportion of damaged grain was observed in pigeon-pea, followed by green gram and black gram with chick-pea the least, both in terms of percentage by weight and percentage by number. Compared to cereals and millets, increase in the true uric content in legumes was higher after a nine-month storage period, and the uric acid level exceeded the safe limit in all the legumes except the chick-pea.

A similar type of study conducted by Vimala (1982) showed that weight losses in four legumes ranged from 1.79 to 5.9 per cent in 12 months of storage, being maximum in green gram and minimum in chickpea, while the true uric acid content ranged from 11.92 to 26.53 mg/100 g. After up to four months of storage, uric acid levels were below 10 mg (fit for human consumption), but after eight months of storage, the content increased to levels (10-20 mg) at which it is unfit for human consumption in all the legumes, except for the chick-pea, in which the level is very close to the permissible range even after 12 months of storage. Coastal area samples of green gram contained the highest concentration of uric acid, namely 31 mg/100 g. Other weight losses reported in legumes are much higher than the two present studies (Venkat Rao et al. 1960; Kapoor et al. 1972; Swaminathan 1977; Adams 1977; Rajyalakshmi 1978). Periodical drying and cleaning of the gram during home storage seems to reduce the damage considerably.

The initial moisture content of the stored legumes ranged between 9 and 10 per cent. There was either a slight increase or decrease (+ 4-14 per cent) in moisture content of legumes stored for nine months. The change in moisture content could have been caused by a change in atmospheric temperature and relative humidity, by periodic sun-drying, by insect infestation, or by a combination of these factors.

A study of the aflatoxin contamination in red gram samples from three agro-climatic regions of Andhra Pradesh showed that no freshly harvested samples were positive for aflatoxin (Rajyalakshmi 1978). After three months of storage, 20.8 per cent of the pigeon-pea samples contained toxin; 4 per cent of these were at a level considered unsafe. After six months the frequency of aflatoxin contamination had further increased.

Carbohydrates

Vimala (1982) has provided further information on changes occurring in carbohydrates in legumes stored in the home for 12 months by rural families of Andhra Pradesh. The starch content of all the legumes decreased as the period of storage increased, but reduction in the starch content was not directly related to infestation. Though insect infestation was highest in green gram, the reduction in starch content after 12 months was maximum in pigeon-pea (7.24 per cent), followed by green gram (6.19 per cent), black gram (3.49 per cent), and chickpea 12.92 per cent).

TABLE. Starch Content (g/100 g) of Fresh and Stored Legumes from Rural Areas of Andhra Pradesh[a]

Legume	Storage period (months)				Percentage of loss		
	0	4	8	12	0-4	0-8	0-12
Green gram (*Phaseolus aureus*)	47.20	45.61	45.18	44.28	3.37	4.28	6.19
Pigeon-pea (*Cajanus cajan*)	41.18	39.79	39.18	38.20	3.38	4.86	7.24
Chick-pea (*Cicer aritinum*)	44.52	43.73	43.44	43.22	1.70	2.43	2.92
Black gram (*Phaseolus mungo*)	34.41	33.81	33.59	33.21	1.74	2.38	3.49

a. Percentage of loss calculated on individual sample basis.

Reducing sugars in four legumes increased at the expense of non-reducing sugars after twelve months of storage, in accordance with the observations of many other investigators (Bottomley et al. 1950; Pingale et al. 1956; Houston et al. 1957; Glass et al. 1959; Lynch et al. 1962; Premvalli et al. 1979; Bhat et al. 1975; Pushpamma and Uma Reddy 1979). The greatest increase in reducing sugars was observed in pigeon-pea 170.22 per cent) followed by green gram (47.75 per cent), chick-pea (34.40 per cent), and black gram (28.87 per cent). This marked increase in reducing sugars could be due to la) changes in moisture, relative humidity, and temperature of the surrounding atmosphere as well as of grains, (b) insect activity, and (c) seed activity. A similar pattern of decrease occurred in non-reducing sugars - from 26.14 to 44.61 per cent in the four legumes.

Amylose content decreased in the four legumes as the storage period increased, but the decrease was not found to be significant (P > .05).

Indigestible residue content of the four legumes increased with the level of insect infestation because of the increase in the husk portion. The greatest percentage increases were observed in green gram (3.65 per cent), followed by pigeon-pea (3.5 per cent), black gram (2.33 per cent), and chick-pea (1.14 per cent). No such increase in indigestible residue content occurred in samples free of infestation.

Decrease in in vitro digestibility of carbohydrates was noticed in all the four legumes. The decrease in digestibility was greatest in green gram (3.87 per cent) followed by pigeon-pea (3.07 per cent), black gram (2.80 per cent), and chick-pea (0.91 per cent). Starch degradation and increase in indigestibte residue content might be the causes of this.

Studies of carbohydrate changes in stored legumes are too scanty for comparison with the present data.

Proteins

A study on home-level storage of legumes for nine months was conducted in rural areas in Andhra Pradesh (Pushpamma and Chittemma Rao 1981). Samples were drawn from families storing legumes under natural conditions and methods of storage. A progressive decrease in protein content occurred during storage of all legumes with a maximum loss in green gram (11 per cent) after nine months of storage. There was an increase in non-protein nitrogen with longer storage, with varying figures for different legumes.

A decrease in free amino nitrogen was seen in both treated and untreated samples in all varieties of legumes during storage. In infested samples this decrease was related to the degree of infestation, showing that free amino acids were utilized for the growth of insects (Uma Reddy 1981).

Uma Reddy (1981) reported a decrease in three amino acids in stored chick-pea, green gram, and pigeonpea. The decrease was higher in lysine than in methionine and tryptophan. Amino acid loss is relatively less in chick-pea than in pigeon-pea and green gram where the degree of insect infestation is high. In green gram, the loss was as high as 50 per cent. Apart from Maillard reactions resulting in the decreased availability of amino acids, the consumption by insect of the germ, which is rich in lysine, could have caused the striking fall in lysine level.

Biological Quality of Proteins

Gradual decrease in the protein biological quality during storage of legumes, as assessed by net protein ratio (NPR) and digestibility coefficient, was reported in the same study.

At the end of one year's storage, only chick-pea retained most of its growth promoting value and digestibility, because it is relatively resistant to insect infestation even when not treated. A significant decrease in NPR and digestibility value was observed during storage of all varieties of both treated and untreated pigeon-pea samples (Uma Reddy and Pushpamma 1980). Similar results have been reported by other researchers (Daniel et al. 1977; Yanni and Zimmerman 1970; Mitchell 1944).

Vitamin Losses

The losses ranged from 13 to 25 per cent for thiamine, 7 to 11 per cent for riboflavin, and 7 to 14 per cent for niacin in all the four legumes after five months of storage, and these increased to 21 to 38 per cent, 14 to 23 per cent, and 20 to 42 per cent after nine months of storage. Of these three, ß-vitamins, riboflavin losses were least in all the legumes.

Thiamine losses were greatest in green gram (42.09 per cent), followed by pigeon-pea (33.01 per cent), black gram (27.71 per cent), and chick-pea (27.14 per cent) in a similar type of study on home-level rural storage of legumes in Andhra Pradesh (Vimala 1982). Similar losses of thiamine and niacin during storage of various food grains have been reported by others (Chitra et al. 1955; Daniel et al. 1977; and Morgan et al. 1945).

Cooking Quality of Stored Legumes

Storage under unfavourable conditions is known to adversely affect legume cooking quality, leading to the hard-to-cook defect, a result both of chemical and enzymatic changes.

The cooking quality of legumes is judged by the amount of water absorbed during soaking as measured by the hydration coefficient, increase in weight and volume after cooking, cooking time, and solids dispersed in the cooked medium.

The cooking time for grain legumes was found to increase with storage. The largest increase in cooking time was observed in stored green gram (52 per cent) followed by black gram (35 per cent), chick-pea (31 per cent), and pigeon-pea (29 per cent). The initial water uptake at a given standard time for legumes also differed. Green gram had the highest water uptake (2.03 9/9) but required the least cooking time, whereas chick-pea showed the minimum water uptake (0.81 9/9) but required the longest cooking time. With storage a progressive decrease in water uptake was observed in ail the legumes Except for the cooking time and water uptake of legumes initially, no relationship was found after storage between the percentage decrease in water uptake and percentage increase in cooking time of legumes (Vimala 1982).

Uma Reddy (1981) believed that the cooking quality of legumes deteriorated on storage: they took a longer time to cook, absorbed less water during soaking, and gave a lower yield after cooking.

CONCLUSION

The information available on grain quality changes, though limited, does definitely indicate that deterioration in the quality of food grains stored for a period of six to twelve months is of a magnitude that deserves the attention of researchers working on the post-harvest problems of food grains. So far reduction in losses was attempted mainly in terms of weight loss . Considering the extent of nutrient losses as well as the lowered biological quality of the grain during storage, attempts to minimize qualitative losses would improve the per capita availability of nutrients to the rural population.

10

Achieving Better Nutrition

Processing of foods involves loss of nutrients. Jo make good such losses, foods are enriched with nutrients. Other staple foods are fortified by governments and other bodies, with the object of overcoming endemic nutritional diseases or as a prophylactic measure. Manufactured consumer goods are also fortified with nutrients, either with the object of ensuring the nutritional adequacy of the product or to achieve better sales. In India several such schemes have been tried with more or less success. Fortification, when undertaken, has to be based on a detailed study of the need and the mode of fortification. Economics, consumer acceptability, and means of achieving the objective should be the guidelines for such a project.

This paper discusses the implications of fortification with minerals, vitamins, amino acids, and proteins. Examples of such fortifications like those of the early experiments of the Central Food Technological Research Institute, Mysore, such as the preparation of synthetic rice using groundnut proteins as a fortificant for protein-poor tapioca flour, preparation of multi-purpose food as a protein concentrate to be used by the consumer as a fortificant of food to provide both vitamins and proteins, and other experiments carried out in Tunisia, Thailand, and Guatemala where wheat flour, rice and corn flour were fortified with vitamins, amino acids and soy flour (in the case of Guatemala) are discussed in detail. The need for fortification and the sociological overtones of fortification are discussed in detail. A striking example of an unproductive type of fortification is shown in the case of baby foods. In the course of a day, a child fed with baby foods and perhaps some weaning foods, receives as much as 109 to 295 per cent of its vitamin requirements. To overcome this, a method has been suggested where vitamin fortification is related to the calorie requirements (or caloric consumption) of the child.

INTRODUCTION

All agricultural raw materials are processed before they can be used as foods. Cereals and pulses are debunked and debranned or polished. Milk is pasteurized or boiled, homogenized, concentrated, or dried. Fruits, if they are to be processed, are expressed, pasteurized, or concentrated. Meat and meat products, if made into processed products, are heat-processed. In all these processing stages, some portion of the nutrients is destroyed or otherwise lost. Some nutrient loss is inevitable, for the raw foods cannot be consumed or preserved.

The term enrichment has been used to signify the act of adding the nutrients that are lost in processing back to the processed food, in such a way that the nutritional value corresponds to that of the unprocessed material.

On the other hand, to overcome certain endemic nutritional diseases or to increase the nutritional quality of staple foods, governments may undertake programmes of providing nutritional ingredients by using staple foods as the medium. Such a process is generally termed fortification (Borenstein 1974). Fortification is also the term used when nutrients are added to new types of foods to make them resemble their natural counterparts. For example, the image to emulate when making synthetic soft drinks is lemon or orange juice, for vegetable milks it is cow milk, and for textured vegetable proteins it is meat or meat products. The terms "enrichment," "fortification," or "nutrification" are often used interchangeably.

ENRICHMENT OF STAPLE FOODS

In order to overcome the losses of vitamins and minerals during milling of wheat, enrichment with these has been practised for the past several decades. In the United States, the Food and Drug Act of 1943 prescribed enrichment of wheat flour with thiamine, riboflavin, niacin, and iron. In the United Kingdom in 1942, food scarcity brought about by the German submarine blockade made the Government enforce the long-extraction order for wheat milling and, to overcome the effect of phytates in bran, enrichment of flour with calcium carbonate (0.15 per cent). This was further increased to 0.3 per cent in 1946 (Sur and Lyengar 1953).

Mandatory enrichment of certain staple foods has now been prescribed in some countries (table 1). In a few other countries recommendations are made for such enrichment. The United States has prescribed enrichment of flour (wheat), self-raising flour, bread, rolls, and buns, and milk and milk products ("Standards of Identity" 1974). The nutrients used for enrichment are the vitamins of the B-group (thiamine, riboflavin, and niacin) and the minerals, calcium and iron.

TABLE. *Enrichment of Cereal Products*

	Per kg weight of flour				
Country	Vitamin B1 mg	Vitamin B2 mg	Niacin mg	Iron mg	Calcium mg
Australia	1.6	2.4	16	14.7	1,000
Canada[a]	4.9	3.0	40	33	1,250
Brezil[b]	4.5	2.5	-	30	1,000
Chile (rice)	6.3	1.3	13	13.3	1,700
Denmark	5.0	5.0	-	30	5,000
Philippinesa[a,b]	5.0	3.0	40	33	1,250
UK[b]	2.4	-	16	16.5	1,250
USA (30 states)[a,b]					
Wheat flour	5.0	3.0	40	33	1,250
Bread	3.2	2.5	28	23	1,200
Rice	6.6	4.0	53	43	1,250

a. Average of the minimum and maximum limits

b. Legally enforced.

Source: Encyclopedia of Food Science, Peterson Johnsons, vol. 3, AVI Publication, Westport, Conn, 1978.

Mandatory enrichment of staple foods is practised only to a limited extent in India. Vanaspati (hydrogenated oil) has to be enriched with vitamin A (25 IU/g).

This has evoked a controversy in recent times, since it has been shown that vitamin A is subject to serious losses, particularly when it is used as a frying medium. Besides, the vitamin A in vanaspati is not of benefit to the population that needs it most, as they may not be able to buy vanaspati (Achaya personal communication 1981).

Mandatory enrichment is also prescribed for margarine (30 IU vitamin A/g) and iodized salt (20 ppm potassium iodide).

Of the vitamins used to enrich wheat flour, it has been reported that 35 per cent of thiamine and 18 to 27 per cent of riboflavin are lost during the processes of cooking or frying used in making doughnuts (Borenstein 1974). Brooke (1968) has reported losses of 50 per cent of added vitamin B1, 39 per cent of vitamin B2, and 40 per cent of niacin.

FORTIFICATION OF FOODS

To overcome a nutritional disease prevalent in a large section of the population, the government or another agency may undertake the supply of the deficient nutritional factor(s) through the medium of food. To do this the missing nutrient will have to be put into a food that the target population is consuming. Many governments have launched such programmes of fortification. Some of the projects have been successful, others have been abandoned mid way.

To be successful, a programme of this nature has to be based on very sound planning. First it has to be ascertained that a need for fortification exists. Nutritional and health surveys to establish the widespread occurrence of a deficiency disease have to precede any such decision. The nutritional diseases that can be prevented by measures of intervention are caused by vitamin and mineral deficiencies. Foods can also be fortified with amino acids or protein concentrates in order to bring about better utilization of the protein. In many nutrition intervention programmes all three components have been combined.

Austin (1979) has considered in detail the factors that govern the implementation of a nutritional intervention programme. They are:

(a) nutritional need, which implies a detailed health and nutrition survey;

(b) commodity system structure, which decides on the food vehicle to be fortified, depending on the food habits, the commodity marketing, and buying habits of the population;

(c) technology for fortification;

(d) consumer acceptability, which probably is the crux of the success of the project ; and, finally,

(e) the economics.

Any programme of fortification will be viable only if all these factors are very carefully surveyed and analysed.

Techniques of Fortification

Techniques of fortification depend on the nature of the medium that is to be fortified, which, in turn, is decided upon after a detailed food habit survey. If it is decided to fortify a staple grain like wheat, a predetermined quantity of the pre-mix containing the nutrients is metered into the mainstream of the flour, and thoroughly mixed.

A similar technique is followed when fortifying milk. The fortificant is added into the mainstream of milk or concentrated milk. Fortification can be done to the dried milk powder also by mixing the fortificant with it.

In all cases, a pre-mix of the fortificant is prepared earlier, so that it provides sufficient fortification when added in the ratio of 0.1 to 0.5 per cent to the bulk.

Fortification of Rice Grains

Rice is eaten as cooked grain so fortification by mixing in fortificants is not possible. Three techniques have been adopted for such grains.

Soaking and drying. A portion of the grains (perhaps 1 per cent of less of the bulk) is soaked in a concentrated solution of vitamins and minerals. The grains are dried and distributed within the bulk in the required proportion. This has two disadvantages. If the target population has not been educated about the fortification programme, it can easily pick out the grains, as they are yellow (due to riboflavin) and discard them. Or, again, rice is usually washed before cooking and in this process the fortificants can wash out.

Spraying. Spraying is similar to the above process but the fortificant is repeatedly sprayed on to the premix, dried and again sprayed till the required concentration is achieved. The final spray usually takes the form of a water-resistant coat to prevent any leaching out of the fortificants (Young 1979).

In this case also the target group might remove the fortified grains. In addition, after cooking, the fortified grain will colour parts of the rest of the rice giving rise to unpleasant yellow patches.

Synthetic grain. The fortificant can be mixed with rice flour and other flours, extruded through a macaroni press with suitable dies. and dried.

The advantage of this process is that it is much simpler than the two earlier processes and can accommodate even a blend of amino acids. However, the grains are again distinctly separate from the other grains and are easily distinguishable from the bulk.

Fortification with Amino Acids and Protein Concentrates

The programmes of fortification of food with vitamins and minerals always stem from a realization that provision of vitamins or minerals can overcome clearly defined deficiencies. Fortification of cereal flours with amino acids or protein concentrates, however, is not based on such definite evidence.

The chemical score of some cereals is shown in table below, which shows that in all of them the limiting amino acid compared to egg protein is lysine, threonine, and tryptophan in that order.

TABLE. Chemical Score of Cereals[a]

Amino acid	Wheat	Rice	Corn	Egg
*Lysine*mg/g nitrogen	170	230	200	440
Chemical score	38.6	52.2	44.5	100
*Threonine*mg/g nitrogen	180	230	280	320
Chemical score	56.3	71.9	87.5	100
*Tryptophan*mg/g nitrogen	70	80	40	90
Chemical score	77.8	88.9	44.4	100

a. Only three amino acids are considered here.

Source: ICMR 1971

Supplementation of cereal diets with lysine or lysine and threonine have always increased the protein efficiency ratio (PER) indicating a better utilization of the rice protein (Jansen 1974). Studies with children have not always given such unequivocal results, though several research workers have reported beneficial effects (Jansen 1974; Pereira et al. 1969). In many of these experiments, the fortified cereal flour has formed only a part of the diet consumed by the children. However, there have also been reports that have not confirmed the supplemental effect with children (Reddy 1971; Albernathy et al. 1972). A detailed discussion of this has been presented in a Protein Advisory Group (PAG) document (1969).

The advantage of fortification of cereals with amino acids is that there is the least amount of change in the nature of the product. The food is not altered in colour or taste. Both L-lysine and L-threonine do not leave any residual taste or flavour.

But addition of dL-methionine brings about a sulphurous taste and flavour when the fortified flours are cooked or heated. This can be avoided by using the hydroxy analogue of methionine, which has the same biological effect as dL-methionine (Protein Advisory Group 1960). With lysine and threonine, toxic effects have not been reported but toxic effects with excess methionine have been reported (Harper et al. 1970; Jansen 1974).

Fortification of cereals with protein concentrates (edible oilseed cakes) has also been advocated. At levels of 8 per cent, Parthasarathy et al. (1964) have shown supplemental effects when de-oiled peanut meal and de-oiled soy meal were added to diets based on ragi (Eleucine coracana) and whole wheat. Such fortification brings about problems of colour, taste, and consumer acceptability. The protein content of the total mix is raised by about 3 to 4 per cent but the content of the limiting amino acid, lysine, is hardly raised with added edible peanut meal.

LARGE-SCALE CONSUMER TRIALS

The US Agency for International Development (USAID) instituted three major field trials for testing out cereal fortification during the period 1970-76. The selected places were Tunisia, where wheat flour used for making bread was fortified; Thailand, where rice grains were fortified; and Guatemala, where corn used for making tortillas was fortified. The nature and levels of fortification are shown in table belo. The investigations were characterized by detailed preliminary nutritional and commodity surveys, technology and feasibility of fortification, and overall effects of fortification. In each place, central fortification as opposed to retail fortification (where the individual consumer is given the option to add the fortificant to the flour when he comes to the mill with the grain for powdering) was also tested. The report of these three investigations provides a detailed study of cereal fortification (Austin 1979). Consumer acceptance of the fortified food in all three places was fairly good. In Thailand there were some difficulties because of the fortificant grain being different from the bulk of rice in colour and taste. But generally it was accepted when the reasons were explained.

TABLE. Fortification of Staple Foods - USAID Projects

	Cornflour (Guatemala)	Wheat flour (Tunisia)	Rice as grain (Thailand)
Fortificants used per tonne of staple food:			
Vitamin B1	21.4 9	8.5 9	5.0 9
Vitamin B2	13.0 9	5.4 9	4.0 9
Niacinamide	154.4 9	63.8 9	-
Vitamin A (10^6 IU/g)	6.2 9	10.0 9	23.7 9
Vitamin D	-	2 x 10(6) IU	-
Tricalcium phosphate	-	50 9	80.0 9
Iron	430 9	55.9 9	
L-Lysine-HCI	1.2 kg	2.0 kg	2.0 kg
L-Threonine	-	-	1.0 kg
Soy flour	6.72 kg	-	-
Cost of grain per tonne ($)	200	200	300
Cost of fortification per tonne of grain ($)	38.628	12.206	69.427
Cost (percentage) of fortification on cost of staple	19.3	6.1	23.14

Source: Austin 1979.

The conclusions drawn from these three field trials were: "The Guatemala study produced very slight but inconclusive evidence of decrease in morbidity or mortality rates, but in Tunisia and Thailand no detectable beneficial effects were observed" (Austin 1979).

INDIAN EXPERIENCE

Indian Multi-purpose Food

One met hod of fortification of the common dietaries is to provide a supplement that the consumer can add to his food to provide the required extra vitamins and proteins. In some ways this is a more elegant approach than broadcast fortification of the staple grains. It is selective in its approach and involves consumer participation. This approach was tried through Indian Multi-purpose Food (IMPF). IMPF is a blend of 75 per cent of edible peanut flour and 25 per cent of Bengal gram (Cicer arietinum). It provided 42 per cent of fairly well-balanced protein with added vitamins, calcium, and iron. It was recommended for use along with other foods and provided a fair amount of the daily requirements. Several studies of its nutritive value both with rats and children were significantly in its favour. Consumer acceptance of IMPF was well established and guidelines for recipes were prepared (Subrahmanyan et al. 1957). The large volume of sales that it enjoyed during the project proved its popularity. However the project was not continued perhaps due to lack of marketing techniques in its sale and poor governmental support (Orr 1972).

TABLE. Chemical Composition of Indian Multi-purpose Food

Constituents	Per 100 g	Constituents	Per 100 g
Protein (g)	41.9	Niacinamide (mg)	14.0
Fat (g)	8.5	Vitamin A (IU)	3.000
Carbohydrates (g)	35.8	Vitamin D (IU)	300
Calories (g)	387.3	Calcium (mg)	660
Thiamine (mg)	1.3	Iron (mg)	5.1
Riboflavin (mg)	3.0		

Sources: V Subrahmanyan et al. 1957

Synthetic Rico

Fortification of a starchy material with a high protein source like edible peanut flour was also tried by the scientists at the Central Food Technological Research Institute (CFTRI) (Subrahmanyan 1954). The technique adopted was to prepare a dough with tapioca flour (Manihot utilissima) and extrude it through a macaroni press with a suitable die and cutter. The grains formed were treated in such a way that they did not disintegrate while cooking but were not too tough for digestion. The grains were fortified with minerals and vitamins by mixing the fortificants with the dough. Nutritional evaluation of the product both with children and rats was carried out. The project was successful both nutritionally and technologically but the programme of establishing a plant at Kerala, which is the major tapioca-growing area, did not come off, for reasons other than the technological feasibility of the project. A similar trial under the sponsorship of USAID was reported to be in progress in 1971 in Thailand, using extruded imitation rice kernels containing lysine, threonine, vitamins, and minerals (Borenstein 1971).

Fortification of Wheat Flour

Fortification of whole wheat flour (atta) with edible peanut flour was recommended by the CFTRI to raise the protein content of wheat flour (normally about 11 to 13.5 per cent) by adding 5 per cent edible peanut flour. Since peanut flour protein is poor in both lysine and sulphur acids, addition of peanut flour to wheat flour did not upgrade the protein of wheat very much. Also it introduced problems of acceptability as the chapati (unleavened bread) made using this flour was dark in colour and not soft. A consumer acceptance trial at Bombay, organized by the Protein Foods and Nutritional Development Association of India, confirmed these findings.

Although the project was initially proposed to be of a fairly large magnitude and was to have been carried out at Poona, it was shifted to Bombay. It involved three fair price shops and was limited to six weeks duration, during July to August 1973. The whole wheat flour (atta) was fortified with 5 per cent edible groundnut flour and some vitamins and minerals, and was sold at the same price as whole wheat flour. Although, according to the questionnaire, 70 per cent were willing to use it, only 8.3 per cent of those who purchased wheat bought the fortified atta. Perhaps it was so because of fear of an unknown food material. The actual number of purchasers of the fortified wheat flour was 174 and the total quantity of the material purchased was only 606 kg. A consumers' survey indicated that only 14 per cent of the users were satisfied, the general complaint being that chapathis made of fortified atta were hard. The fortified flour, although packaged in polythene bags, showed significant insect infestation and this contributed greatly to non-acceptance of the product. [Aiyar personal communication 1982].

Fortification of Bread

Modern Bakeries, a public sector organization of the Government of India, opted for fortification of its Modern bread with lysine, vitamins, and a few minerals. The objective was obviously to correct the lysine deficiency in wheat flour and make the bread better utilized metabolically. Significant improvement in the weight and height of children fed chapati using lysine-fortified wheat has been reported by Shashi Prabha and Bagchi (1978). Similar reports using bread from lysine-fortified wheat flour has been reported in Haitian children by King et al. (1963). Lysine in bread does not alter the taste or texture of bread but probably part of it becomes unavailable during baking if reducing sugars are present. However the fortification of Modern bread with lysine has now been given up by the producers. Occasionally they use soya flour as a wheat supplement, particularly for bread used in school feeding programmes.

Fortification of Salt

Iron-deficiency anaemia is prevalent among women of child-bearing age and in children, both in developed and developing countries (WHO 1975). It is said to be the most widespread nutritional deficiency disease in India (Gopalan 1970). Parasitic hookworm infection has been reported to be a major cause of this. The disease has been shown to contribute significantly to the mortality and morbidity of children, and of women of child-bearing age. One of the remedies suggested is to provide ferrous salt with folic acid to pregnant and lactating mothers and young children. This can benefit only those with whom contact can be established, and does not help the much larger groups in villages and poorly developed areas. It appears to have met with little success (Nadiger et al. 1980). The disease is so widespread that only a broadcast method like fortification of a basic food item consumed by all groups irrespective of age, economic status, and rural or urban setting can meet the requirements. Such a vehicle for fortification in the Indian milieu is salt. A project for salt fortification was undertaken by the Food and Nutrition Board of the Government of India in collaboration with the National Institute of Nutrition (NIN).

In fortification of any food material with iron salts the major consideration is the absorption and utilization of iron by the human body. The Workshop on Measures to Increase Iron in Foods and Diets, organized by the Food and Nutrition Board of the National Academy of Science, United States, in 1970, observed that, "in large-scale studies iron fortification has failed to show expected benefits because iron in the compound used for fortification has been unavailable, or the amount of iron added has been insignificant" (National Academy of Sciences 1970).

TABLE 5. Bio-availability of Iron Salts Used in Fortification

Vehicle	Iron compound	Other additives	Bio-availability
Salt	Ferrous sulphate	None	Poor
	Ferrous sulphate	Ascorbic acid	Increased
	Ferric phosphate	None	Poor
	Ferric phosphate	Ascorbic acid	Increased
	Ferrous sulphate	Sodium hexa meta phosphate and sodium acid sulphate	Poor
	Ferrous sulphate	Orthophosphoric acid	Decreased on storage
	Ferrous sulphate	Sodium acid sulphate	Good
Sugar	Ferrous sulphate	None	Poor
	Ferrous sulphate	Ascorbic acid	Increased
Tea	Ferric chloride	None	Poor
MSG[a]	Ferrous sulphate	Zinc stearate and vitamin A	Good

a. Monosodium glutarnate

Source: Zoller et al. 1980.

The characteristics of some of the iron salts used for fortification and their absorption in human systems is shown in table 5. Elwood et al. (1970) using labelled iron in chapati have shown poor absorption of both iron in the wheat and from the iron salt used for fortification. They have stressed the need to discover or develop an iron salt that can be added to flour in relatively high concentration but that will not affect the keeping quality or baking quality of wheat.

To overcome the problems of bio-availability, storage, and utilization in Indian culinary practices, the fortification of salt with iron has been intensively studied by the NIN (NIN 1977). Bio-availability of different combinations of labled ferrous or ferric salts with orthophosphoric acid or its salts and acid sodium sulphate have been studied (Narasinga Rao and Vijayasarathy 1975; Narasinga Rao 1978; Nadiger et al. 1980). The problems of storage in containers that are usually used in Indian homes have been investigated, and the availability of the iron in different forms when used with salt has been tested ("Standards of Identity" 1974). The final formula recommended by the NIN was to fortify the salt with a combination of ferric orthophosphate and sodium acid sulphate, at such a level as to contain 1 mg of iron per gram of salt. The cost of such fortification was estimated to be 17 paise per kg.

A large quantity of this fortified salt was produced by the Food and Nutrition Board and the consumer acceptance and the nutritional effect on consumers was tested at four centres - the All-India Institute of Hygiene and Public Health, Calcutta; the All-India Institute of Medical Sciences, New Delhi; the National Institute of Nutrition, Hyderabad; and the Institute of Child Health, Madras. The first three were tested under rural conditions and the last under urban conditions. The results have confirmed that salt fortified with iron is acceptable and that the consumption of such salt was associated with an improvement in haemoglobin status and a reduction in prevalence of anaemia even in the presence of hookworm. The programme is now ready for adoption on a large scale (Dutta personal communication 1981).

Iron Fortification of Other Foods

Fortification of other foods with iron has been tried. Fortification of sugar has been studied in detail by Disler et al. (1975) and Layrisse et al. (1976a, b), and has proved a suitable vehicle for iron fortification. However, sugar may not be a very useful medium for fortification in India as many of the lower economic groups use very little sugar. Fortification of tea was seriously considered in India by the Food and Nutrition Board as a vehicle for iron fortification. Unlike coffee, tea is consumed throughout the country and even by children. However, the bio-availability of iron in tea is much reduced by the chemical constituents of tea.

An interesting development is the fortification of monosodium glutamate (MSG). MSG is used regularly in many of the South-East Asian countries, and almost 94 per cent of the population in the Philippines use MSG. Tests with ferrous sulphate encapsulated with zinc stearate and with ferric phosphate showed that the former had an availability coefficient of 70 per cent and the latter 50 per cent as compared to 100 per cent for ferrous sulphate. Both were free-flowing and did not impart any significant colour (Zoller et al. 1980).

Fortification of Processed Foods

TABLE. Vitamin and Mineral Intake of Children on Fortified Baby Foods[a]

	Intake/day	Requirement	Percentage Requirement
Vitamin A (IU)	2,430	1,333	182 3
Vitamin D (IU)	590	200	295 0
Vitamin B1 (mg)	0.91	0 6	151.7
Vitamin B2 (mg)	1.39	0.8	1738
Vitamin B6 (mg)	0 62	-	-
Niacinamide (mg)	9.14	8.0	114.2
Calcium (mg)	1.62	0.4	405.0
Iron (mg)	7.64	7.0	109.0

a. A child of six months weighing 6 kg requires about 700 calories/day. Assuming that the child is receiving 2/3 its calories from baby food and 1/3 calories from weaning food, namely, 467 calories from baby food and 233 calories from weaning food, the child will get the above quantities of vitamins against its daily requirement.

Many processed foods are fortified with vitamins, calcium, and iron, and in some cases with amino acids. Infant foods and weaning foods are usually fortified either on the initiative of the manufacturers or in accordance with the specifications of the Indian Standards Institution (ISI). Added to this there are vitamin preparations with mineral salts (calcium and iron) which are being advertised as a "must" for children, and which doctors suggest that mothers should give their children. Possibilities of over-vitaminization or over-feeding with minerals for children may happen in this way. To avoid this, Lachance (1970) has suggested that the vitaminization of a food should be based on its calorie content, that is, the vitamin or minerals should be proportionate to the calories that the food supplies.

In addition, the mother is advised by doctors to give tablets of calcium with vitamin D (160 IU), tablets of vitamin C /200 mg) and vitamins A and D preparations (6.000 IU. vitamin A and 1,000 IU vitamin D) or multi-vitamin tablets

But any programme that is subsidized has a short life. Hence unless the consumer is convinced that the fortification is for the general good and is willing to pay for the benefit, the programme cannot continue. The prevailing costs of vitamins and minerals are shown in table 10. The costs of amino acids (1977 prices) are also shown (Austin 1979).

Fortification of processed foods with vitamins is comparatively cheap and because of this there has been indiscriminate fortification. This can be reduced if the suggestion of linking calories and vitamins in the products is adopted.

IS FOOD FORTIFICATION NECESSARY?

The question as to whether food fortification is necessary has to be considered under three subheads: (a) fortification of staple foods with the object of prophylaxis against general malnutrition; (b) fortification of staple foods with the object of wiping out a nutritional disease that is prevalent; and (c) fortification of processed foods.

Fortification as a Means of Improving Nutrition

There have been many fortification programmes that governments have undertaken. Such programmes have been based on the fortification of an individual staple food with vitamins and minerals. Most of these programmes have had very little popular support. Given the option, the consumers have preferred the unfortified product to the fortified one, though there was no price difference between the two (Austin 1979; Aiyar personal communication 1981).

This is generally so because the public has an inherent dislike of "new foods," particularly so when the good that the food is supposed to do is not perceptible, as in the case of fortification with vitamins or amino acids. Past experience has been that, when possible, the consumer picks out the fortificant grains from the bulk. These diets, which are within the means of a large majority of the population, can meet all their requirements. Fortification of any one staple food becomes unnecessary if the consumers are educated in proper dietaries.

TABLE. Balanced Diets as Recommended by the ICMR

Food item	Adult		Children	
	Man	Woman	1-3 years	4 6 years
	(g/day)		(g/day)	
Cereals	520	440	175	270
Pulses	50	45	35	35
Leafy vegetables	40	100	40	50
Other. vegetables	70	40	20	30
Roots and tubers	60	50	10	20
Milk	200	150	300	250
Oil or fat	45	25	75	25
Sugar or jaggery	35	20	30	40
Approximate cost per diet (Rs)	3.13	2.43	1.95	2.21

Source: ICMR 1981.

Parboiling that is practised in several parts of the country is a natural and accepted form of fortification of cereal grains with vitamins. Parboiled rice has a higher concentration of vitamins than raw or polished rice.

Similar techniques for wheat (termed bulgar wheat) are also available. Consumer acceptance of such processes is assured. Similarly a high extraction of flour in roller flour milling ensures better vitamin content than low extraction. A third method that will not have problems of consumer acceptance is genetic improvement of cereals to evolve strains with better nutritional content.

Amino acid fortification of staple food is an entirely different question. Experiments based on animals, or even on selected groups of children, cannot answer the question of the necessity for fortification. The benefit of fortification shown in some trials has been negated by other groups, and no unequivocal answer regarding benefits is yet available. Hegsted's dissenting note to the 1969 PAG meeting on fortification is still valid: "If the benefit to be derived is unknown, cost/benefit analysis is not possible. However, programmes of this nature based upon theoretical or laboratory data alone are fraught with danger." The cost of fortification of cereals with amino acids is heavy, particularly if the second limiting amino acid, threonine, has also to be included.

Fortification with protein concentrates has some advantages over mere fortification with amino acids: it provides calories in addition to the required amino acids, and it will not increase the cost of the blend since protein concentrates sell at nearly the same price as cereals. Soy flour has the advantage over peanut flour in providing a sufficient quantity of lysine. Technological improvements, however, are necessary to ensure that other disadvantages, like undesirable flavours and colours, or changes in texture of the processed products using such flours, are eliminated. Consumer acceptance is a necessary component in such fortification. A project for a tailored tur dhal to overcome the legume shortage is being considered. This will have the advantage of providing a balanced protein with necessary calories and vitamins for that large section of the population who may not be in a position to purchase costly pulses. Technological possibilities to achieve consumer acceptance have to be explored.

Fortification of Staple Foods to Overcome Specific Nutritional Diseases

Programmes like salt fortification with iron or iodine, or fortification of milk powder with vitamin A, have much to recommend them. Such programmes have a specific purpose and can achieve the objective in a definite period. Such programmes are of immense value in wiping out nutritional diseases.

Fortification of Processed Foods

It has now become common practice to fortify all processed foods with vitamins, minerals, and even amino acids. The need for such fortification is frequently not established, and it is often used as a gimmick for sales promotion. It has been pointed out earlier that fortification is always disproportionate to the requirements. Excess intake of vitamins and minerals may bring about serious nutritional problems. The suggestion to link the calorie level of such foods with vitamin fortification is a very useful one.

CONCLUSION

Processing of foods involves loss of nutrients. To make good such losses foods are enriched with nutrients. Staple foods are fortified, by governments or other bodies, with the object of overcoming endemic nutritional diseases or as a prophylactic measure. Manufactured consumer goods are also fortified with nutrients, either with the object of ensuring the nutritional adequacy of the product or to achieve better sales. In India several such schemes have been tried with more or less success. Fortification when undertaken has to be based on a detailed study of the need, and the mode of fortification. Economics, consumer acceptability, and means of achieving the objective should be the guidelines for such projects.

11

Distributional Aspects of Caloric Consumption: Implications for Food Policy

A major thrust of this chapter is to develop a quantitative model for linking nutritional intake (calories) with agricultural production and prices. The population has been stratified by regions (rural and urban), and expenditure classes and demand models for calories have been estimated for each group. The redistributional effect on calorie intake has been quantified under alternative assumptions concerning the supply. The effects of fluctuations in agricultural production are quantified by linking land holding distribution with income distribution. Implications of utilization of agricultural surplus are traced.

Although enough food items are now produced to provide every person with an acceptable level of caloric consumption, nearly half the population suffers from caloric deprivation, and this is alarmingly large for the bottom strata (Dandekar and Rath 1971; Rao and Vivekananda 1979). In fact, in periods of upswings in grain production, stocks accumulate even before the food deficiencies of the poor are met. The real problem is that hunger is overwhelmingly the result of income inequality and the lack of purchasing power among the lower strata of the population. Viewed thus, the food problem would appear to be mostly a result of distribution, and the solution to it would not merely lie in producing more food.

The paper analyses the calorie intakes of various strata. It examines (a) the extent of caloric deprivation among the bottom classes; (b) the effects of income and prices on calorie intake;

and (c) the implications of the results for policy priorities. Since distribution is one of the determining factors of caloric deprivation among the poor, the analysis is focused on expenditure classes. We present in the next section the basic model employed and in the last section the results.

Murty and Radhakrishna (1981) have provided, for India, parameter estimates for the Nasse model computed from the data on consumer expenditure provided by the National Sample Survey Organization (NSSO) for the rounds starting from number 2 (1952/53) to number 25 (1970/71). Six of the nine commodity groups considered by the model belong to food, and the model has been estimated separately for five expenditure classes in rural and for five in urban areas. Some of the details of the expenditure classes are given in table below.

The following procedure has been adopted in working out the calories contained in one rupee worth of expenditure on each of the six food commodity groups. For 1961/62, NSSO provides quantities consumed and their values at a fairly desegregate commodity level for a number of expenditure classes. These quantities have been converted into calories, and then the calories contained in one ropes worth of expenditure on each commodity group is estimated. The expenditures are at 1961/62 prices, and using commodity group-specific price deflators, the calories contained in one rupee worth of expenditure at 1970/71 prices have been estimated.

These estimates exhibit a few broad features of interest. First, the cereal group is the cheapest source of obtaining calories, and the meat, eggs, and fish group the costliest. Roughly, a rapee expended on cereals yields the same quantity of calories as that of Rs 2-2.5 expended on items belonging to the "other food" group, and Rs 5-10 on meat, fish, and eggs. Second, a rural consumer derives more calories from one rupee spent on any commodity group than his counterpart in urban areas. Third, calories contained in one rupee worth of expenditure on any commodity group decline as one moves from bottom to top classes. For instance, from a ropee spent on cereals, the bottom classes derive 4,410 kcal and 3,780 kcal respectively in rural and urban areas, while the corresponding figures for the top classes are respectively 3,330 and 2,730 kcal. Thus within each commodity group, the upper strata replaces calorie-intensive items by items with a lower calorie content.

Table below shows the shares of various commodity groups at mean level, which is obtained by relating the quantities estimated from the demand models to the results given in table 2. The cereal group is the major source of calories for all classes both in rural and urban areas. It accounts for 87 per cent of the calorie intake by the bottom rural class and 67 per cent for the rural top class, while the percentages for corresponding urban classes are 77 and 26. the "other food" group is also an important source of calories for the higher classes, particularly in urban areas. Only in the top urban class does one notice some diversification in the sources of calories; otherwise, cereal and other food groups account for most of the calorie intake.

TABLE. Percentage Distribution of Calorie Intake by Sources in Bottom and Top Classes

	Percentage distribution of calorie intake			
Sources	**Rural bottom class (class 1)**	**Rural top class (class 5)**	**Urban bottom class (class 1)**	**Urban top class (class 5)**
Cereals	87	67	77	26
Milk and milk products	1	7	1	12
Edible oils	1	3	3	11
Meat, fish, and eggs	1	1	1	2
Sugar and gur	2	7	3	10
Other food	8	15	15	39

Making use of parameter estimates of the demand models and the results given in table above, calorie consumption function, i.e. equation (4), has been estimated for each expenditure class. Per capita calorie intake during 1970/71 has computed at mean level from the calorie consumption function separately for each expenditure class. The per capita per diem calorie intake works out to be 2,423 kcal, for rural areas, 2,239 kcal for urban areas, and 2,394 kcal for India as a whole. Clearly, though at the aggregate level there is no caloric deficiency, deprivation is observed for the bottom two rural and bottom three urban classes if we take the norm as 2,250 kcal. The extent of deprivation is around 35 per cent for the bottommost rural class and 44 per cent for the bottommost urban class. Since the expenditure classes considered by the demand models are broad, and the population distribution is also not uniform for rural and urban areas, neither the percentage of population affected by calorie deficiency, nor rural-urban comparisons of the intensity of deprivation, could be deduced. However, an earlier study has shown that the percentage of population affected by calorie deficiency is higher in urban than rural areas (Rao and Vivekananda 1979).

Table below provides the average propensity to consume calories by size of classes obtained by dividing the mean calorie intake by the corresponding mean expenditure. It can be seen that the average propensity to consume calories vary markedly across the classes. For instance, it works out to be 3,044 kcal for bottommost rural class and 1,228 kcal for the topmost class, while for the bottom and top urban classes it works out to be 2,519 and 769 kcal respectively. The low propensities observed for higher classes is obviously due to their higher propensity to consume non-calorie food items.

TABLE. Average and Marginal Propensities to Consume Calories (expressed in kcal per rupee)

	Rural		Urban	
Expenditure classes	**Average propensity to consume calories**	**Marginal propensity to consume calories**	**Average propensity to consume calories**	**Marginal propensity to consume calories**
1	3,044	2,820	2,519	2,550
2	2,576	2,250	2,029	1,740
3	2,136	1,380	1,729	1,230
4	1,813	1,080	1,354	840
5	1,228	510	769	330

The marginal propensity to consume calories computed at mean levels varies substantially across the classes; the low values obtained for rural and urban top classes are striking (table 4). The caloric elasticity with respect to total expenditure (TIC) also exhibits a more or less similar pattern. The caloric elasticity is unity for bottom classes implying that calorie intake expands proportionately with total expenditure; and it declines to about half for top classes implying that a 1 per cent increase in per capita expenditure results in a half per cent increase in calorie intake. Among price effects, cereal price has a strong influence on calorie intake and its magnitude declines as we move from lower to higher groups. For instance, a 1 per cent rise in cereal price will result in a cut of 0.8 and 0.7 per cent in the calorie intake by the bottom rural and bottom urban classes respectively and the corresponding percentages for the top classes are only 0.2 and 0.1. Thus, per capita expenditure and cereal price mostly determine the calorie intake level. Only in the case of upper classes does price level of the "other food" group have some influence on calorie intake.

POLICY IMPLICATIONS

It does seem that the total quantity of food utilized for domestic consumption can provide the minimum calorie requirement. Despite this, a large section of the population is deprived of the minimum. The real problem appears to be the distributional imbalance in the cereal consumption, since cereal is the major source of calorie intake. It is likely that it will continue to be a major source for some time to come at least for the lower strata.

What then are the policies to alleviate the caloric deprivation of the poor? The redistribution strategy of limiting the consumption of the upper income groups and transferring the purchasing power to the poor has received wide attention in Indian plan formulations. Simple calculations using the caloric elasticity would show that in order to eradicate caloric deficiency, it would be necessary to increase the per capita expenditure by 54 and 14 per cent for the bottom first and second rural classes respectively, and by 78 per cent, 48, and 13 per cent for the bottom first to third urban classes respectively, through slicing the expenditures of the higher income groups.

However, such consumption transfers would result in additional demand for cereals, since the cereal group takes a comparatively greater share in the marginal rupee spent by the lower classes. If the excess demand for cereals is not eased through supply expansion, cereal price tends to rise, and the impact of the price rise would be heaviest on the poorest sections of the population not covered by purchasing power transfers (see Murty and Radhakrishna 1981).

It is thus clear that meeting the caloric deficiency by consumption transfers presupposes stepping up of the aggregate cereal consumption. On the other hand, a number of schemes have been proposed for nationwide public distribution that are workable with the existing supply. The implementation of the schemes is still a hope. It must be pointed out that public distribution as tried earlier was essentially designed to serve through rationing the requirements of the cities and industrial centres that cover also the rich class. The rural areas and their vulnerable sections had little access to the public distribution system. Such inadequate coverage would be worse than no rationing at all, since the vulnerable population outside the rationed areas may then be without assured supplies.

The National Dairy Development Board's Project on Oilseed Growers' Cooperatives

Edible oils constitute the most important source of calories in the Indian diet, after food grains. However, oilseed production has stagnated, and India - once an oilseed exporter - now imports 1 million tonnes of edible oils per year Oilseed crops are grown on marginal land with low inputs if any; processing is inefficient; and marketing is characterized by high profit margins and speculative practices that exploit the growers. To combat these problems, the Government of India asked the National Dairy Development Board to develop the edible oil industry along the lines of the highly successful milk cooperatives. Growers' federations have been formed in four Indian states to supply agricultural inputs, secure stable product prices, and organize marketing channels. Under this project, production, processing, and marketing are handled wholly by the growers' organizations. Benefits are already showing - in Gujarat the grower members received the highest price to date for groundnut, in Madhya Pradesh, for soybean. It is anticipated that by the seventh year the growers' co-operatives will include 350, 000 oilseed growers in 8.000 villages of India.

INTRODUCTION

After food grains, edible oils constitute the most important source of calories in the Indian diet. However, the per capita consumption of fats and oils in the Indian diet is less than half the assessed nutritional requirement. Even though a sizeable area (about 16 million hectares) is under oilseed cultivation, the per hectare production is extremely low and stagnant, resulting in dependence on the import of over one million tonnes of edible oil every year to meet the country's growing demand. Grown on marginal lands under rain-fed conditions, without improved technology, unsupported by proper and timely credit-input-extension facilities, and faced with highly unstable

The views expressed in this article are the author's own and do not necessarily reflect the views of the organization with which he is associated. product prices, oilseeds in India are considered to be "orphan" crops. Oilseed processing is highly inefficient, with per unit production costs considerably higher than the world average. It has also failed to tap the high protein content of oilseeds, which could help farmers to obtain stable, remunerative returns, and consumers a quality product.

BACKGROUND

The major oilseeds grown in India are the groundnut, rapeseed and mustard, sesame, castor, linseed, sunflower, and soybean. Among the edible oilseeds, groundnut uses for nearly 65 per cent of the area under oilseed cultivation constitute nearly half of the indigeneous edible oil production. This is followed by rapeseed-mustard, which accounts for nearly 30 per cent of the area. In the post-Green Revolution period, yields of rapeseed-mustard improved, but groundnut yields have experienced no such improvement except in irrigated areas and in certain tracts where improved farm practices are applied. During the planning period, the overall groundnut production increased slightly above 2 per cent per year but progress was uneven. For example, the Saurashtra region of Gujarat State recorded a rise of about 5 per cent, but in vast tracts of Maharashtra, Karnataka, and North Gujarat, production has declined in recent years.

Highly unstable yields and highly fluctuating prices are the crux of the groundnut problem. The unstable trends in groundnut yields are due to its total dependence on rainfall, which itself is erratic both spatially and temporally. This in turn makes the returns from groundnut low, very risky, and highly volatile. Therefore, the farmer does not invest in much-needed inputs such as improved seeds, seed treatment chemicals, rhizobium culture, pesticides, fungicides, and fertilizers, etc., which are necessary to raise the per hectare yield. To encourage the farmer to invest in these inputs, it is necessary not only to create awareness of their benefits, but to make all of them available at the right time, in appropriate quantities and at a price that is within the farmer's reach. The farmer also needs to be assured of reasonable prices for the produce. This can be realized only if the oilseed processing and marketing of oilseed products are done by the farmers themselves.

Speculative marketing practices and high profit margins characterize the industry. The traders and other intermediaries have a powerful hold in the oilseeds and oil trade, both in the domestic as well as export markets. They have contributed to the high degree of instability in the edible oil market through hoarding, illegal forward trading, and speculation. The speculative operations of the private trade has deleterious effects on the incentive of the farmers during the years of good production and on the welfare of consumers in periods of shortage. Oilseeds in general are grown as a rain-fed crop. resulting in marked seasonality in arrival of the crop in the markets. The bulk of the crop is received in the mandis (local markets) during October-November, when the traders, by collective action, depress prices.

The price decreases all the more if there is a bumper crop, as in 1975/76. They tend to avoid purchases till the farmers are ready to accept prices dictated by them. The farmers are unable to withhold their produce as they need cash immediately after harvest, and are at the mercy of the hardened traders. The reward for increasing production is often ruin. The intermediaries take double advantage of the seasonality of oilseed crops when they artificially jack up prices, especially of edible oil, in the off-season period, and particularly when the oncoming year is not expected to be a normal one. Thus the marketing of oilseeds and their products is subject to a high degree of exploitation, and is self-degenerating. These problems are too complex to be solved by ad hoc measures, and call for an integrated approach towards restructuring the entire edible oil sector - oilseed production, processing, and marketing - so that growers are not robbed of their returns, and consumers are saved from exploitation.

THE ANAND MODEL

A similar situation, though of lesser intensity, prevailed in the case of milk about 30 years ago, when the exploitation by middlemen had forced the milk producers to sell milk at throw-away prices. The answer to such a situation was found in developing a system that enabled milk producers to own and operate not only their processing facility, but also the inputs, delivery machinery, and the marketing of finished products.

The integrated approach first proved very successful in Kaira - the AMUL organization - and through the Operation Flood programme, which aimed to establish similar farmers' organizations in 18 districts in the country. The programme when implemented did face some preliminary problems, but these were overcome, so much so that when the International Bank for Reconstruction and Development world Bank) came forward to finance dairy projects in Karnataka, Rajasthan, and Madhya Pradesh, they saw to it that similar farmers' organizations were created in these states.

The result was that 12 milk producers' co-operative unions were established. Based on the success of these 30 unions, bodies owned by the farmers themselves, the other phase of dairy development, Operation Flood II, is being implemented to bring the total number of such district unions to 155, and cover some 10 million farmers.

The success of these 30 unions have further demonstrated that problems can best be solved if the agency implementing the programme is owned and operated by the producers themselves (only then can it be sensitive to their needs), who employ the best professionals to carry out their policies. It has also proved that the responsibility of such an organization is not only to guarantee a remunerative market for the rurally produced commodity (milk in this case) by its member producers, but also guarantees a package of inputs for enhancing production (milk) and ploughs back the profit for further development.

Again, in the case of milk, since marketing of the product (especially milk for city supplies) was largely controlled by middlemen/traders/cattle keepers, etc., who were capable of using any means, fair or foul, the organized dairy was not in a position to take a commanding share of the market and thereby pay a better price to the milk producers.

Market intervention was therefore required to enable these dairies to supply milk to the consumers at reasonable prices. This was the first time that food aid - skimmed milk powder, and butter oil in this case - was used to prime the pump of the modern dairies in four metropolitan cities (Calcutta, Bombay, Delhi, and Madras), and permit them to meet the demand on a much larger scale. The funds generated through the sale of skimmed milk powder and butter oil were used to build up farmers' organizations in the milksheds that supplied these metropolitan cities.

At about the same time, the Government was viewing the situation in the edible oil industry very seriously. Wide fluctuations in the prices of edible oils and oilcakes had produced repercussions in the dairy sector and thus on the total availability of fats and oils in the Indian diet - and had also resulted in stagnancy in the production of oilseeds.

The question was: could the experience gained in milk be utilized for the production and distribution of oils and oilseeds? Although the nature of production and the commodity itself was quite different from milk, other situations, such as the lack of bargaining power of the producers and the organized actions of the traders/middlemen, allowed the situation in question to continue in favour of the latter.

An integrated approach to the entire situation indicated that the experience gained by the National Dairy Development Board (NDDB) in organizing milk producers, and using food aid to prime the pump as well as generate funds for further investments in the programme, could very well be utilized to restructure the oilseed/oil industry as well.

THE PROJECT

At the request of the Government of India, the NDDB formulated a project

"Restructuring of Oil and Oilseed Production and Marketing" in October 1977 - again using food aid (edible oils) to generate funds for use in agricultural development in the oilseed sector - with the following objectives

a. to co-ordinate the modernization of oilseed production, processing, and the marketing of vegetable oils, oilseeds, and by-products, in areas covered by the project;

b. to procure and market imported and indigenously produced vegetable oils in such a way as would contribute to the stabilization of supplies at levels that will be fair to consumer and growers;

c. to increase the opportunities for productive and remunerative employment in the selected major oilseeds growing areas;

d. to generate the funds required for the establishment of a modernized oilseeds and vegetable oil industry based on oilseed growers' co-operatives, which will put the functions of oilseeds processing and marketing into the producers own hands;

e. to devise and implement a programme of investment and development that will enable growers to increase their oilseeds production, and their returns, while also increasing the efficiency of the processing and marketing functions, through the growers' own co-operatives.

It was expected that oilseeds production in the areas covered by the project would increase by about 30 per cent by the end of the project period.

THE SCOPE OF THE PROJECT

The project spreads over a seven-year period with a total outlay of Rs 1,50û million, and has two phases. The first phase of the project is of three years, and will cover eight contiguous areas, each having approximately 250,000 hectares under oilseed cultivation. At the end of phase 1, the progress made under the project will be reviewed, and subject to availability of additional funds, etc., the project may be extended to cover another twelve contiguous areas in major oilseeds growing States.

At the end of the project period, it is expected that the project will cover 350,000 oilseeds growers, organized in about 8,000 villages, located in the states of Gujarat, Madhya Pradesh, Andhra Pradesh, Tamilnadu, Maharashtra, and Orissa.

The requisite funds (Rs 1,500 million) would be generated through the sale of 250,000 tonnes of donated edible oils received during the seven-year period. The funds generated from the sale of these commodities would be invested on the following action items:

Action Items	**Investment (Rs in million)**
Creation of processing facilities	596.75
Operations research and CIS studies	18.50
Market research and testing	22.95
Oilseeds Growers' Co-operative Organization and Production Enhancement	439.74
Product and Process Development	11.70
Manpower Development	30.00
Project implementation	102.87
TOTAL	1,222.51

The remaining Rs 277.49 million would be used as a sinking fund for Federation share capital, procurement support, operational losses, management training, and co-operative development.

The project was approved by the Government of India in September 1978. It was launched in Gujarat in July 1979, in Madhya Pradesh in October 1979, in Andhra Pradesh in October 1980, and in Tamilnadu in January 1981. Maharashtra and Orissa were also to be covered by the project by mid-1982.

The project envisages the setting up of village-level primary oilseed growers' co-operative societies, affiliated to a state-level co-operative oilseed growers' federation. This farmers' organization will not only own and operate the facilities for procurement, processing, and marketing of oilseeds and their products, but will also have ties for provision of technical inputs for increased production of oilseeds (mainly through rise in the current per hectare yield in the project area).

The project envisages that the grower members, through their elected Boards of producers, will control and manage all activities right from production to processing with the help of hired professionals so that the producers' interests are represented throughout the hierarchy.

All the functions that the traders and other intermediaries were performing will be taken over by the farmers' own organization and in the interests of farmers themselves.

The by-laws of the village level oilseed growers' co-operative societies and the state-level federations have been drawn in such a manner as to ensure participation of only genuine oilseed growers in the running of the co-operatives.

One-third of the members of the Board retire each year by rotation, thereby providing an opportunity for entry of new leadership through a democratic process.

In the event that a member does not supply a specified minimum quantity of oilseeds to the village cooperative or engages himself directly or indirectly in trading activities, he loses voting rights. Continuous and concurrent audit is yet another feature that would make such co-operatives more aware of the need to follow strict financial practices in their day-to-day operations.

The State-level Co-operative Oilseeds Growers' Federations, created as implementing agencies under the project, have two major components for investment.

Creation of Processing Facilities

The funds allocated, Rs 596.75 million, or about 40 per cent of the total generated funds, would be spent on hiring existing processing facilities and renovating them if necessary, and on creating modern processing, storage, movement, and packing facilities. The money would be advanced to the project implementing agency - state-level federations - on a 30 per cent grant and 70 per cent loan basis. It is expected that by the seventh year of the project, farmer-owned oilseed processing units would handle some 1.5 million tonnes of oilseeds.

Growers' Co-operative Organization and Oilseed Production Enhancement Programmes

The funds allocated, Rs 439.74 million, or about 29 per cent of the total generated funds, would be made available to the Co-operative Oilseeds Growers' Federations to enable them to undertake a number of activities at the micro and macro levels of project operations.

CO-OPERATIVE ORGANIZATION AND SUPERVISION

The first step in the production enhancement process will be to organize the oilseed growers into village level co-operative societies.

A promotional and educational effort will be undertaken by a mobile team, which normally consists of technicians, a co-operative specialist, agricultural extension workers, etc. Generally, a mobile team will be required to look after 20 villages.

It will hold meetings with villagers and exhibit films showing how milk producers have been able to form and operate their "Anand Pattern" co-operatives.

Eventually, two or more receptive farmers will be selected to participate in the NDDB's Farmers' Induction Programme (which has already trained over 15,000 farmers in the last decade). These farmers will be brought to Anand or any other similar training centre for about two to three days to see for themselves how "Anand Pattern" co-operatives function, and what benefits they provide to their members.

Normally, a Village-level Co-operative Oilseeds Growers' Society will cover an area of approximately 800 to 1,000 hectares. Initially, societies will be assisted by the NDDB mobile team in organizing their business, and in their record-keeping system (which consists of trading, financial, and operational activities).

A local resident is hired by the society to be its paid manager (called secretary).

The project provides an initial two-year subsidy for expenses on salaries of the secretary and assistant, and rent of the office building, to help it become viable relatively quickly. The Village-level Oilseeds Growers' Cooperative Society undertakes procurement of oilseeds from grower members based on quality and also provides the much needed infrastructure that will enable the oilseed growers' federation to implement the production enhancement programme.

ARRANGING PRODUCTION ENHANCEMENT INPUTS

An effort will be made to mobilize the resources already available from various agencies, especially the Government. The purpose is to avoid duplication in manpower and financial investments. In fact, most of the inputs available are not properly utilized by the farmers for various reasons.

The aim is therefore to assure maximum utilization of the available resources, laying a strong foundation for future utilization of the inputs envisaged in the project at the optimum level. The oilseed growers' co-operatives will arrange supply of improved seed, rhizobium culture, fungicides for seed treatment, fertilizer, insecticides, and other inputs on cash payment basis or in kind. Where such facilities do not exist, mobile team members will assist oilseed growers to obtain finance from other sources for input purchases.

DEMONSTRATION PLOTS

In each village or in each co-operative society, one or more village co operative representative will be selected to serve as a demonstration oilseed grower. Two demonstration plots of four hectares each will be established in each village to demonstrate the advantages of improved oilseed cultivation techniques.

The plots will be periodically supervised and inspected by the agricultural specialists of the mobile team. During the growing season, "field-day" demonstration programmes will be organized at the demonstration plots to introduce co-operative members to innovative practices and test results. The grower-demonstrator will be given a subsidy (in various forms) of up to approximately 25 per cent of the total cost of production or Rs 500 per hectare.

The produce from these demonstration plots, if meeting the prescribed field standard, will be procured as seed material at an incentive price of, say, 5 to 10 per cent over the market price. After cleaning, grading, and testing for physical purity and germination, these seeds will be supplied to the members of the society.

MANAGEMENT SUBSIDIES AND EQUIPMENT/STORAGE GRANT

Each organized co-operative society, during its first two years of operation, will be provided an organizational and managerial subsidy, equivalent to Rs 3,000/- per year. This is to assist the society in hiring a manager (secretary), and in meeting start-up costs. Each co-operative also receives equipment, such as a large hanging bar scale for weighing the oilseed produce, a moisture meter, a small balance for testing produce to calculate shelling percentage, refraction, and impurity discounts.

This provision has been made mainly to help ensure the viability of village-level co-operatives in their initial years of operations, when the oilseeds procured through these societies may not be adequate to generate incomes, earned in the form of society commission from the federation, to meet their running costs. Grants to develop local storage facilities are also given to the co-operatives.

DISTRICT FARMS

In each contiguous area of 250,000 hectares, where project implementation will be initiated, the project authority will seek to acquire a farm of 300-400 hectares (to be donated by the state government) to serve the purpose of a district farm. In the absence of such large farms, the project envisages having a cluster of smaller farms belonging to rural community councils (panchayats), State Department of Agriculture, Agricultural High Schools, etc., to be operated on a joint management basis.

These farms are expected not only to demonstrate improved cultivation methods and ways of dealing with local cultivation problems, but also to serve as training bases for farmers and extension workers serving on the mobile teams. Thus they would play an important role in the multiplication of foundation seeds.

AREA AGRONOMIC CENTRES

It is planned to establish one area agronomic centre during phase-1 of the project and one to two additional centres during the third to seventh years of the project.

The Area Agronomic Centres will serve several purposes. First, they will help in applied production research by conducting local trials of oilseeds technology developed at regional and possibly international agronomic research stations. Second, they will produce foundation seeds for multiplication by the district farms. Third, they will disseminate recent research findings in oilseeds to project participants. Land for these centres is to be provided by the state governments. Each centre will have a total proposed land area of about 500 hectares.

With a view to crush the produce in units owned by the members and to gain some experience in processing and marketing, two processing units acquired under the project are functioning at Bhavnagar and Jamnagar in Gujarat.

In addition, the Gujarat federation has also obtained an industrial licence to set up a groundnut processing plant of 500 tonnes per day capacity in the Saurashtra region of Junagadh district and a 200 tonnes per day capacity mustard processing plant in north Gujarat.

The Madhya Pradesh federation has obtained a letter of intent for setting up a 300 tonnes per day capacity soybean processing plant at Ujjain. Both Gujarat and Madhya Pradesh federations have started marketing oil and vanaspati.

Procurement operations have increased considerably, and the farmers have gained in terms of better returns. In Madhya Pradesh, the State-level Oilseeds Growers' Federation procured 5,000 tonnes of soybeans in the 1980/81 season, and were expected to procure around 15,000 tonnes during the 1981/82 season. Similarly, in Gujarat, as against 3,000 tonnes of groundnut procured in 1979/80, procurement reached a figure of 21,000 tonnes in 1980/81 through organized co-operative societies.

Besides, the Gujarat federation procured 182,000 tonnes of groundnuts as a part of the joint venture with the Gujarat State Co-operative Marketing Federation in 1980/81. The total turn-over of Gujarat and Madhya Pradesh federations for 1980/81 was around Rs 700 million and it was expected to be around Rs 800 million during 1981/82. The prices paid to oilseeds growers averaged Rs 4200/- per tonne for groundnut, offered by the Gujarat federation, and Rs 2,760/- per tonne for soybean, offered by the Madhya Pradesh federation. This has benefited the growers, particularly as they received such prices at their doorstep.

12

Meeting Dietary Fat Requirements

Fat is an expensive commodity, and average consumption over several years in India has been just 12 to 14 grams per day.

Dietary surveys reveal low intakes in Karnataka and Tamilnadu, average intakes in Maharashtra and Andhra Pradesh, and high intakes in Gujarat and Kerala. Urban dwellers consume more than do rural, and metropolitan cities have comparatively high in takes. Income is a major determinant of fat in take, but regional culinary practices play a part.

Average diets in various parts of India carry-16 to 24 grams of invisible fat, which is present at some level in practically every foodstuff. Half such latent fat derives from the staple food. Total fat intakes from these diets come to 36 to 44 grams per day, contributing 13 to 27 (average 18) per cent to dietary calories. Even poor diets (1,000 to 1,300 calories) still carry 10 grams or 9 energy per cent as fat,

Apart from their other dietary and nutritional functions, fats carry the two essential fatty acids (EFAs), linoleic and linolenic. These act in the body as precursors of the prostaglandins, a group of some 20 compounds that are synthesized in various organs and perform a large number of vital functions. FAO/VVHO recommend that EFA-derived calories should constitute, as percentages in various diets; adults, 3, pregnant women, 4.5, and nursing mothers, 7. Indian diets, if adequate in calories, would easily meet these norms; even average diets, below par in calories, carry 3.5 to 4.5 energy per cent as EFAs.

High proportions, from 35 to 45 per cent, of fat-derived calories in the diets of advanced countries, together with the living stresses of modem life, appear to conduce to coronary heart disease. For India moderate fat levels of 18 to 20 per cent appear optimal. The actual range is from 8 to 32 per cent, and at all energy intakes, the invisible fat contribution is close to 9 energy per cent. Average Indian diets, 20 to 30 per cent energy-deficient, have the recommended fat level. Consumption of 2,800 calories by any adult, including a pregnant or lactating woman, would call for 30 grams of external fat.

Infants on breast milk receive 20 to 27 grams of fat daily. This quantity halves in the first year, while external fat intake varies with both region and income. Breast-milk substitutes should ensure 50 to 60 energy per cent as fat and weaning foods 25. EFA levels in infant foods should be 3 to 5 per cent (with a further 1 per cent of higher unsaturated fatty acids if possible) and in weaning foods 3 per cent.

Vegetable oil production in India has stagnated at around 9 million tonnes over the last decade, and prices have sky-rocketed. Any breakthroughs in respect of the groundnut which constitutes ha/f the total oilseed crop, would make the strongest impact. The sunflower, safflower, and niger all have potential to raise oil production, as does further exploitation of ricebran and of existing tree oilseeds. Soya and the oil palm are of very limited utility.

ACTUAL FAT INTAKES IN INDIA

Developing countries are characterized by low intakes of fat, since of all basic foodstuffs, fat is one of the most expensive. Figures for consumption of fat in the various states of India are to be had from various dietary surveys, and these show quite wide variations (Protein Foods Association of India 1969 and 1972; Hindustan Thompson Associates 1972).

States like Karnataka and Tamilnadu consume low quantities of fat (9 and 10 grams per day per head), Maharashtra and Andhra Pradesh are medium consumers (11 and 14 grams), and Gujarat and Kerala are high consumers (18 and 21 grams). Urban dwellers consume more per head than do rural, and metropolitan cities have the highest intake of all: Calcutta city dwellers, for example, average 26 grams daily.

Income is a major determinant regulating use. As income goes up, so does, almost proportionately, the consumption of fat. Thus in Calcutta city both poor and rich spend 6-7 per cent of their income on edible oil. But cultural factors also seem to play a part. Thus the proportion of families who use no fat at all ranges from low figures of 0 to 5 per cent of all families in Maharashtra, Gujarat, and Tamilnadu to as high as 43-60 per cent in Karnataka, Kerala, and Andhra Pradesh. Income differences cannot explain these very great disparities in fat usage. Nor will a factor like vegetarianism, because in Gujarat, which is 69 per cent vegetarian, practically all families buy and use fat, and in Tamilnadu, which is only 16 per cent vegetarian, nearly all families use fat. The most likely explanation is that regional culinary practices are reflected in fat usage patterns, for example, a Bengali will probably strive for at least a few drops of mustard oil to flavour his food.

Aboriginals represent a special group. Dietary surveys show that daily calorie intakes can range from as high as 2,600 to just half that figure, but fat intakes per se are always below 5 grams per day, and frequently non-existent (Achaya 1978a).

So far we have been talking of fat that is used for cooking or at table. But almost all foodstuffs contain some amount of fat. Oilseeds and nuts, milk, meat, fish, and eggs are obvious examples, and rice, wheat, pulses, grains, tapioca, tubers, and vegetables are less apparent ones.

Detailed dietary surveys that list individual foodstuffs permit the quantities of such invisible or unseen fats to be calculated. Such calculations for average diets in the north, south, east, and west of India show that some 16 to 24 grams of fat are present invisibly in the food items (other than fat per se) that are consumed. Nearly half of this comes from the staple foods, usually cereals and tubers (such as tapioca in Kerala) since these are eaten in amounts of over 400 grams per day. If the visible fat consumption is taken into account, the total fat intake in some regions varies from 36 to 44 grams per day. This would make a contribution to calorie intake of between

13 and 27, average 18 per cent, which is not a really low figure. It is also noteworthy that even very poor diets, containing no visible fat and furnishing only 1,000-1,300 calories daily, will still carry about 10 grams of invisible fat, making a calorie contribution of about 9 per cent. Many poor tribal diets would fall into this category (Achaya 1976, 1979b).

QUALITY OF FAT CONSUMED

Until fairly recently, fats were viewed in nutritional terms as concentrated sources of calories, as carriers of fat-soluble vitamins and as adding palatability to the diet.

A little later the role of linoleic acid as an essential fatty acid was revealed, first in preventing skin scaliness, particularly in children; then as a protective factor in stress situations, and thereafter as a dietary factor that appeared to be implicated in preventing the deposition of plaques of cholesterol esters on the aorta walls, a condition termed atherosclerosis. In the last two decades, the importance of dietary linoleic and even linolenic acid has been dramatically highlighted following a realization of their role as precursors of the prostaglandins.

These compounds are synthesized from the two essential fatty acids (EFA), and there are twenty or more prostaglandins, all closely related in structure. Prostaglandins perform a variety of vital functions such as muscle contraction, blood platelet aggregation, relief of pain, fertility in the male, conception and reproduction in the female, and many others.

Prostaglandins appear to be synthesized by many organs in the body as and when required, and the quantities present at any time are reckoned in milligrams.

The quantity of dietary linoleic acid required for synthesis of the requisite quantity of prostaglandins is not known with certainty, and has to be deduced by the dose of linoleic acid required to correct certain skin conditions, and for restoration of certain biochemical indices. An FAO/WHO Expert Consultation on Dietary Fats held in October 1977, after reviewing the evidence available, felt that 3 per cent of the energy level in adult diets should derive from dietary EFA (FAO/WHO 1977).

The total visible and invisible fat intakes of average diets in the four cardinal areas of the country furnish a linoleic acid content that ranges from 17 to 28 per cent of the fat present, four-fifths coming from the invisible fat component. In terms of weight this works out to 7 to 10 grams of linoleic acid daily, or 3.5 to 4.5 per cent. For the upper income groups, the figure is 11 to 15 grams (about 6 energy per cent), and for the low-income groups 6 to 11 grams, or 3.6 to 5.7 (average 4.5) energy per cent (Achaya 1979c).

COMPARISON WITH RECOMMENDED LEVELS

In various parts of the world fats furnish very varying proportions of the total calories. In developing countries, the figure is 35 to 45 per cent, and, since calories levels are also high, the quantity of fat exceeds 100 grams daily. Eskimos derive over 50 per cent of their energy from fat, without apparent evidence of degenerative disease. Developing countries show wide variations. In India very low-income and deprived groups, with inadequate intakes of 1,300 to 1,500 calories, get all their fat invisibly, and this constitutes some 8 energy per cent. At the other extreme affluent groups in many regions average 70 grams of total fat a day, constituting 32 energy per cent, and many individuals may go even higher. Constant intake of diets that are high in total calories and in fats, as now prevalent in most advanced countries, taken together with the stresses of modern living, appear to be associated with increasing risk of coronary heart disease. More moderate levels, say, of about 18-20 energy per cent as fat, were suggested by FAO/WHO for developing countries with existing low intakes. The Indian Council of Medical Research in their 1981 recommendations suggest 20 energy per cent for adults, about half of which is expected to come from invisible fats (Indian Council of Medical Research 1981).

At all income and consumption levels in India, the contribution of invisible fats is almost exactly 9 energy per cent. Thus a visible fat consumption that will furnish an equal 9-10 energy per cent is called for. For an average adult male optimally requiring 2,800 calories this would mean about 30 grams of visible fat per se a day. For average diets as they stand at present, with calorie intakes that are about 20 to 30 per cent below par, the fat contribution to calories now averages 18 per cent, and no increase in fat consumption appears to be called for. Pregnant and lactating women , provided they consume the levels of calories recommended, would also get enough fat.

What about fat quality? The quantity of 3 energy per cent of essential fatty acids believed to be adequate is met by average Indian diets; these carry 3.5 to 4.5 energy per cent of EFA, because of the fortunate circumstance that the invisible fats in cereals and pulses are rich in this component. If calorie and fat intakes are both raised, it would seem prudent to use unsaturated oils rich in EFA like safflower, sunflower, sesame, and groundnut.

Needs of EFA during pregnancy go up from 3 to 4.5 per cent, and during lactation to 5 to 7 per cent. Again, use of unsaturated oils is advisable, apart from extra food as a whole.

FAT INTAKES IN CHILDHOOD

How do Indian children fare in terms of fat? Not too badly at first, but not too well as they grow older. When breast milk is available in adequate amounts, in the first year of life an Indian baby gets 20 to 27 grams of fat daily, and in the second year 12 to 16 grams, from this source alone. Thereafter both income and region begin to play their part. In a state characterized by a generally high fat consumption such as Gujarat, daily fat intakes are between 16 and 17 grams in the third, fourth, and fifth years of life. In a "low-fat" state such as Karnataka the figures lie between only 5 and 12 grams. High-income groups in Calcutta city, who can afford to buy animal milk, show an average fat intake of about 25 grams a day for these three post-weaning years, against just 9 grams for low-income groups in the same city. However, since breast feeding is less prevalent among the affluent, their children may sometimes paradoxically show lower fat intakes in the early years than the lowest-income, breast-feeding segment (Achaya 1979).

Thus on grounds of fat intake, as for many other reasons, breast feeding is undoubtedly the best feeding method. Artificial foods that substitute for breast milk should be so formulated as to ensure the high level of 50 to 60 energy per cent through fat that is required by infants. By the time of weaning the fat requirement and tolerance drop sharply, and weaning diets require about 25 energy per cent as fat, or about 17 grams daily. Weaning food formulations must take this into account, as should supplementary weaning foods that are made in the home.

In terms of essential fatty acids, children on breast milk during their first year obtain from this source 2 grams per day (or 3.6 energy per cent) of linoleic acid and one gram per day (or 1.8 energy per cent) of other poly-unsaturated fatty acids. In the second year these figures drop to 1.5 to 2.5 grams per day (2 to 3 energy per cent), and 0.4 to 0.7 grams per day (0.4 to 0.8 energy per cent). By the fifth year the actual linoleic acid intake drops to just 1.5 to 2.8 energy per cent, after which the levels gradually move up to the adult pattern of 3.5 to 4.5 energy per cent (Achaya 1979). Thus weaning and the pre-school years mark a low point in respect of EFA intake, as they do in other nutritional respects. Fats provide energy without undue bulk, and the use of unsaturated liquid oils in larger amounts in weaning foods, though costly, seems to be called for.

EDIBLE FAT SHORTAGES

Edible cooking oil or hydrogenated fat gets costlier all the time. This price rise has been disproportionate to that of other foods. In the decade 1965-1975, while the price index of foodstuffs as a whole rose by a factor of about 2.6, that of edible fats increased nearly 3.5 times. Apart from general inflation, which affected the price even of foodstuffs produced in abundance, the shortage of edible oils in the country pushed up their prices abnormally (Chandrasekaran and Achaya 1980).

What is the extent of this shortage? Production of all vegetable oils in the country has hovered at 2.0 to 2.5 million tonnes over the last decade, with an average annual increase of just about 1.2 per cent. On the other hand the demand for oils has gradually been rising, partly because of the increase in population by 2.5 per cent annually, and partly because a rising standard of living of the upper-income groups of society is always reflected in an increase in consumption both of edible oils, as well as of products like soap and paints that are made from vegetable oils. As a result we have been forced to import from the world market an increasingly larger quantity of oils. As international prices have risen the amount being spent for such buying from abroad has now reached a figure of some Rs 800 crores annually. This is true even when the cheapest oils on the world market are imported, such as palm oil from Malaysia, rape/mustard oil from Canada, soybean oil from the United States, and sunflower oil from the USSR.

The shortfall is well recognized, and each five-year plan has set targets that have unfortunately never been realized. While it may be true that emphasis on wheat and rice production has caused a movement away from growing crops like pulses and oilseeds, there are a few problems specific to the latter. The groundnut is our major oilseed and accounts for half the total vegetable oil in India. Hence any increase in yield through genetic engineering would be of vital importance. As a cross-pollinated crop, hybrid breeding is difficult to effect. Using seed irradiation, mutants have been developed with several plus points such as large seed size, high yield per hectare, and high oil contents, which could serve to bring about a Green Revolution in this important oilseed. A constant fear with the groundnut is the aflatoxin problem to which the oilcake is prone, which has led to some reluctance in markets abroad to use this cake as animal feed. Varieties innately resistant to the growth of the mould have been identified, but there are other regional eco-agricultural parameters that must be maintained as well.

New oilseeds have been suggested that might improve the oil position. The sunflower has been developed as a very rich source of oil, especially in the USSR and East European countries. The sunflower is a cross-pollinated crop that requires that each individual floret in a head is pollinated through extensive bee activity, and there have been disappointments with poor seed filling when the crop has been grown in southern India.

Cross-pollination has been achieved by labourers with cloth gloves on their hands walking between the rows brushing pollen from flower to flower, and some self-pollinating varieties have also been developed. The sunflower has a deep tap root and is thus much less affected by drought than is the shallow-growing groundnut; this is important since drought years always result in low groundnut yields. The spectacutar success in Malaysia of palm oil and palm kernel oil (a rare case of two oils from the same fruit, one from the flesh and another from the seed kernel, each with different characteristics and end-uses) is unlikely to solve India's edible oil problem, since climatic factors would confine any palm oil plantations to Kerala, the Little Andaman isle and just a few other hot, humid areas. The soybean has been much publicized, but in my view does not really suit India's requirements as an oilseed crop. It is a poor source of oil (only 16-18 per cent), and after processing gives rise to a huge quantity of protein cake. This will have to go into processed protein foods for which markets are small and lie in the upper economic brackets who are not short of protein in their diets. Thus soybean oilcake will end up as an Indian export item, which is hardly the best use of our scarce cultivated land area (Achaya 1975).

As the gap between availability and market demand widens, the prices rise, the consumer suffers and the Government is pressed to use hard-earned foreign exchange in ever greater amounts to import oil. The matter is further complicated by the very speculative nature of the oil market, in which moneyed persons control the buying and selling of both oilseeds and oils, and can create critical shortages so as to push up prices and then sell at a profit. Accordingly the National Dairy Development Board, at the request of the Government, is carrying forward a plan to use vegetable oil, donated from abroad in the first instance, to stabilize market prices. At the same time this would generate the money required to modernize and restructure the vegetable oil sector through co-operative production and marketing of oilseeds and oil, to a point where it would have its say in the oilseed economy (National Dairy Development Board 1980).

Fat is an expensive commodity, and average intakes in India are low. Income is a major determinant, but regional culinary practices play their part.

Average Indian diets carry substantial quantities of invisible fat, half of which derives from the staple food. Total fat intakes average about 18 per cent of dietary energy, and just meet the FAO/WHO recommended level (18 to 20 per cent) for developing countries. Indian average diets are energy deficient, and an adult level of 2,800 calories would call for doubling the quantity of external fat consumed. Norms prescribed for essential fatty acid levels are easily met by Indian diets, both actual and recommended.

Infants derive a large proportion of their energy from the fat in breast milk. Infant and weaning foods should ensure adequate fat and essential fatty acid levels.

Vegetable oil production in India has been stagnant for a decade. Breakthroughs in groundnut technology would have a large impact. Sunflower, safflower, and niger have potential, but hardly soya or the oil palm.

COMPOSITE PROGRAMME FOR WOMEN AND PRE-SCHOOL CHILDREN

The Composite Programme for Women and Pre-school Children (CPWPC), an upgraded pre-school feeding programme, aims at improving the nutritional status of young children through provision of a package of services. The programme is jointly undertakrn by the Rural Development Department of Kerala, Mahila Samajams, and CAPE.

The recently constructed balwadi building is the focal point for delivery of services. Each CPWPC unit envisages provision of services to 100 children and 40 expectant and nursing women. As of July 1981, 2,161 CPWPC units have been planned.

The programme plans for a food supplement of 312 calories and 12 grams of protein for each enrolled participant. Medical services are provided by the staff of a nearby health centre. The balwadi teacher is expected to impart child-care education to mothers and non-formal education to children.

Assistance is provided for purchasing 10 goats for 10 families, 10 chickens for each of 10 families, and for developing a kitchen garden to 20 families.

Programme evaluation indicates that the majority of enrolled participants are from the lower socioeconomic group. Almost all interviewed mothers indicated that they received the food supplement regularly. Mothers recalled the names of two to three vaccines received by the child. Child-care education sessions have been attended by 75 per cent of the mothers. More than 80 per cent of the mothers participated in nutrition education camps. About 30 per cent of the mothers reported establishing kitchen gardens and an equal percentage of mothers had established poultry units. Lambs were received by 24 per cent of the families.

The successful delivery of services to the intended group is attributed to the mode of operation and response to the changing situation.

Literally speaking there are thousands of voluntary agencies, both big and small, involved in various fields of human endeavour at international, national, regional, state, and village levels. In India, though the Government has taken on the task of conducting child nutrition programmes, many international and national agencies provide the assistance necessary for their implementation. The predominant voluntary agencies assisting the Government efforts are Mahila Mandals, Youth Clubs, Child Welfare Councils/Societies, Catholic Relief Society, and CARE. CARE, an international relief and developmental organization, has been providing assistance since 1950 and functions within the framework of Indian Government policy and programmes. Although relief during emergencies remained the base of CAR E's activities during its first decade in India, its attention is increasingly focused on nutrition and health programmes. In recent years, CARE has been encouraging and assisting local voluntary agencies in the development and implementation of nutrition programmes at the grass-roots levels.

The core of this paper is drawn from a case-study of an integrated nutrition health programme in Kerala jointly undertaken by the Rural Development Department of Kerala, Mahila Samajams, and CARE.

THE COMPOSITE PROGRAMME (CPWPC)

The CPWPC programme has been in operation since 1975 and aims at transforming vertical pre-school feeding, health, pre-school education, and applied nutrition programmes into an integrated maternal child-care programme.

Through provision of a package of services comprising the food supplement, non-formal education and medical care for children, and child-care education for mothers, the programme aims at improving the nutritional status of young children.

Mahila Samajams have the responsibility of implementing the programme at the village level. CARE provides assistance in the form of food, funding for visual aids for child-care education, equipment for regular monitoring, and evaluation of the programme.

Each CPWPC unit envisages provision of services to 100 pre-school children and 30-40 expectant nursing mothers. As of July 1981, 2,161 CPWPC units had been established. An evaluation component is built into the design of the programme.

In 1979, after four years of implementation, a large-scale field survey was conducted to study the efficiency of the delivery of services, the level of programme participation, and the utilization of services, and to measure the impact of the programme on the nutritional status of the participants.

PROGRAMME PARTICIPANTS

The selection of children on the basis of economic and nutritional status is defined clearly and strictly adhered to. The president of the Mahila Samajam and the medical staff of the nearby health centre select the programme participants.

Almost all the participants are below six years of age, and 13 per cent of them are less than two years of age. Except for a few, the participants are from traditionally lower castes and under-privileged groups. The source of income of the majority of participant families is labour.

TABLE. Age of Programme Participants on the Survey Day

Age (in months)	Children (%)	Age (in months)	Children (%)
0-6	0.1	37-48	33.5
7-12	1.5	49-60	25.4
13-24	11.7	61-72	1.5
25-36	26.1	73 and more	0.2
		TOTAL	100

TABLE. Caste of the Programme Participants

Caste	Percentage	Caste	Percentage
Upper castes	1	Non-Hindu-Muslims	9
Lower castes	42	Non-Hindu-Christians	14
Scheduled caste	17	Others	14
Scheduled tribes	3	TOTAL	100

TABLE. Occupation of Father of Programme Participants

Occupation	Fathers (%)	Occupation	Fathers (%)
Labourer	51.5	Teacher/Clerk	3.7
Traditional service	16.0	Business trade	11.1
Tenant farmer	0.1	Others	11.3
Cultivator	6.6	TOTAL	100.3

PROGRAMME INPUTS

Balwadi Building

At each CPWPC site a balwadi building is constructed with financial assistance from Mahila Samajams, CARE, and the Government of Kerala.

The cost of construction at present is Rs 12,000 (US$1,500). Nearly 30-40 per cent of the funds for land, building material, and construction are generated by the Mahila Samajams, who have legal ownership of the building. The balwadi is the venue for distribution of the nutrition supplement, dispensing of preventive and curative medical services, and imparting child-care education to mothers and non-formal education to children. In short, these structures are the focal point of integrated maternal child health activities in the village. This building has imparted a real sense of performance to the existing ill-housed programmes that were conducted under the shadow of trees or in someone's house.

Food Supplement

The programme is designed to provide an enrolled child a cooked meal of 312 calories and 12 grams of protein for 300 days per year. Children attending the balwadi classes receive an additional meal of 100 calories and 3 grams of protein.

The data suggest that the food is distributed regularly at the centre. The response of mothers to questions on the adequacy and taste of the centre meal, and the purpose of the feeding programme indicates a positive attitude towards the programme. The vast majority of mothers agreed that the taste of the meal was acceptable to the children. Also, the mothers felt that the food given at the balwadi was adequate. More than 90 per cent of the mothers regarded the objective of providing a meal as "improvement of health" of the children.

TABLE. Attitudes of Mothers towards the Programme

Attitude	Mothers (%)
Quantum of food	
Adequate	93 4
Not adequate	5.9
Do not know	0 9
Regularity of the programme	
Yes	99.0
No	1.0
Taste of food	
Like	98 4
Not liked	1 5
Justification for the programme	
Poverty	1 7
Hunger	2.2
Health	91.1
Incentive for attending balwadi	1.4
Do not know	4.3

In analysing the data it was found that a child attending the balwadi received a meal of 404 calories and 22 grams of protein. A meal of 320 calories and 17 grams of protein is given to non-balwadi participants. The entire quantity of food, according to 73 per cent of the mothers, is fed to the enrolled child. Another 24 per cent reported feeding between "half" and "most" of the food the participant child.

TABLE. Quantity of Centre-food Reported to be Fed to the Programme Child

Quantity of centre-food	Children (%)
All	73.1
Most	15 7
Half	8.8
Less than half	1.9
Other	0.2
TOTAL	99 7

Medical Services

Both preventive and curative medical care is to be provided to programme beneficiaries by a medical doctor once a month and an auxiliary nurse mid-wife (ANM) once a week.

TABLE. Utilization of Health Services

Type of service	Mothers (%)
Medicine for illness	7.8
Vitamin A	52.0
Multi-vitamins	41.0
Deworming medicine	20.0
Immunization:	
DPT	52.5
BCG	38.9
Polio	49.0

Evidence collected suggests that on an average a child received protection against two-three childhood diseases such as diphtheria, whooping cough, pertusis, polio, and tuberculosis. More than half the mothers recalled receiving prophylactic doses of vitamin A for their enrolled child. Also, mothers of one-fifth of the children mentioned receiving preventive doses against worms for the participant child. Although the majority of children experienced two-three episodes of illness in a year, less than 10 per cent of the children were treated for illness at the balwadi centre.

Child-care Education

Important concepts concerning certain practices that contribute to child malnutrition, morbidity, and mortality are communicated to mothers of the enrolled children through fortnightly sessions at the balwadi centre, home visits to participants homes, and six monthly camps of two days' duration. The balwadi teacher, trained for a period of one month, has the responsibility of imparting child-care education. For effective communication, posters and pamphlets have been developed. As a tool for personal instructions to the mothers about the health and nutritional requirement of the young child, a coloured growth chart is provided for each programme child.

TABLE. Frequency of Attendance at Child-care Sessions

Frequency	Mothers (%)
Never	26.9
Once a week	2.6
Once a fortnight	2.0
Once a month	16.6
Once in 2 months	4.7
Once in 3 months	16.5
Whenever they have time	31.4

About 73 per cent of the interviewed mothers reported participating in the fortnightly education sessions. Nearly 22 per cent of the mothers attended the sessions once a month. Another 40 per cent participated in the educational activity whenever they found time. The majority of the 27 per cent of mothers who never attended the education sessions mentioned "time" as a constraint for their participation. Nutrition camps were attended by about 80 per cent of the mothers, and about 74 per cent attended more than once.

TABLE. Frequency of Attendance in Child-care Camps

Number of camps	Mothers(%)
0	18.2
1-2	38.2
3-4	35.0
4 and more	7.5
Do not know	2.0

The growth chart used for monitoring the body weight of the child was seen by 72 per cent of the mothers, and 70 per cent of them understood what it conveyed. in a similar programme in Pondicherry, although the children were weighed regularly, the growth charts were seen by less than one-fourth of the participants' mothers. Data on awareness and knowledge of selected child-care practices indicated that a large number of mothers grasped the messages that were being conveyed in the education sessions.

Economic Activities

Mahila Samajams are encouraged to organize economic activities to increase the income of mothers. Assistance is planned for 20 mothers per balwadi in a year to develop a kitchen/backyard garden. Ten families of the participant children are given 10 chickens each on the condition that they provide 100 eggs to the Mahila Samajam. These eggs in turn are supplied to 5 families for hatching and rearing. Provision is made for supplying goats to 10 families. These families are to return a female lamb to the Mahila Samajam for supply to other mothers.

TABLE. Economic Activities Undertaken by Mahila Samajams

Type of activity	Families reported/Receiving assistance (%)
Kitchen garden	30
Poultry units	30
Goats	26

Evidently 30 per cent of the families have established a garden in their backyard. Goats and poultry units have been received by 26 and 30 per cent of the families respectively.

TABLE. Comparison of Perceived Advantages by Mothers and the Purpose of Their Visit to the Balwadi

Services	Mothers	
	Purpose of visit(%)	Perceived advantages (%)
Collection of food	79.4	95.0
Immunization	17.0	31.4
Medical relief services	9.1	12.4
Medicine	10.6	22.8
Vitamin A drops	5.3	12.8
Iron and folic acid	2 7	7.3
Deworming medicine	2.5	4.7
Education	49.7	46.0

Besides these three types of activity, some of the Mahila Samajams are organizing radio assembly units. tailoring classes, and dairy units.

CONCLUSIONS

In evaluating the programme efforts, it was observed that the implementers are able to transfer the project plan to an operationally viable programme. The efficiency of delivery of most of the services is high. Also, the programme has been successful in reaching young children, especially in the age-group of six months to three years. And, having reached the intended group, the services are well utilized. The families of participant children were not only beneficiaries but also partners in the programme implementation. In this programme the participant families are not treated as dependants but encouraged to help themselves and others from the community.

The CPWPC programme has been successful in bringing together interests and resources of different agencies. The key to successful programme operation is freedom in the mode of operation. Also, the implementers were in a capacity to experiment with innovative approaches even at the risk of failure.

DISCUSSION

The Food Mix

The relatively smaller increase over the last 25 years in the price of fine cereals like rice and wheat, compared to those both of coarse cereals like jowar and bajra and of pulses, has adversely affected the nutrition of the poor. Even at a 1,500-calorie-per-day level, the data show a willingness to buy more costly foods, and at 1,800 calories, a shift to purchase non-food items is noticeable. Thus simply the provision of a larger income may not bring about better nutrition, and one may have to consider other measures to correct this tendency, either through the agency of the ration system or by payment of wages in kind.

In drawing up health cards for use in the CARE project in Kerala, red has been shown as the zone of danger. Yet in the Poshak study in Madhya Pradesh, red was perceived by local mothers as a sign of health and well-being.

Fats in the Diet

What would be the critical minimum requirement of external fat? In practice the answer was probably nil, since invisible fat in Indian cereal diets is substantial and, fortunately, high in essential fatty acids as well. Indeed many people, perhaps a third of the Indian population, eat no external fat at all. Consumption of just 400 grams of any Indian cereal alone would furnish at least 10 grams of latent fat.

Of course addition of fat would raise the calorie density of children's foods, but the cost of this world be high. The soybean is important in agriculture. especially in Madhya Pradesh, but has little role as a food that would help the poor as it cannot be consumed directly. On the other hand, when processed the oil would of course find use, but the protein cake must necessarily be incorporated into processed foods, which then go out of reach of the needy.

13

Nutrition: Better of Achieving

1. Agriculture, nutrition, and food science have seldom been discussed together.
2. Agricultural production would continue to remain short in relation to overall nutritional requirements.
3. Cost of production of most of the agricultural produce is higher than the purchasing power of masses.
4. In spite of the fact that there exists a very large fund of knowledge for bringing about technological changes, the overall adoption of new technologies that involve agricultural produce has been slow in spite of commitment to accept the new technologies as in national interests.
5. With the growth rate of population in India, there would have to be continuous effort in improving the overall food production.
6. The lab-to-land programme has many examples of feasible adoption of high productivity. There should be a plan that would attempt to attain overall agricultural production according to nutritional needs.
7. The question arises as to whether it would not have been better to continue encouraging the use of fertilizer and help the farmer in achieving a higher rate of production. Similarly, in case of other inputs like pesticides, cost of power, and fuel for irrigation, the policy should have been to minimize the cost of these inputs.

8. Prospects for India becoming a net exporter of agricultural produce rather than depending upon any imports are good.
9. Many forms of incentives have been designed for industrial development and for export, but in the area of agriculture such incentives are very few. In reality there is negative approach to higher agricultural production as the urbanites and policy-makers continue to scoff at any successes made in agriculture.
10. There is much less interaction between agriculture and government than between industry and government.
11. Poverty is growing in rural areas. Urban areas are receiving more poor people from rural areas every year.
12. It might be possible to evolve a long-term policy where the farmer would be assured of a price that is reasonable and has linkage with any escalation.
13. Linkages of credit and productivity would enable the farmer to use effectively more inputs and improved technology.
14. There is a need to organize well better credit linkages for land development, irrigation and other needs like post-harvest facilities of storage, handling, and marketing.
15. If full benefit of production is to be taken, extension of credit for intermediate steps has to be considered, as this would protect many of the commodities from deterioration in terms of quality, of nutrition, and of damage to the product.
16. A country that is predominantly rural with the obvious lack of rural development suffers due to the lack of infrastructure facilities for the proper marketing of agricultural commodities.
17. Agro-industries are on the threshold of an expansion programme. However, it is unfortunate that most of the agriculture commodities for industries are not produced according to industry requirements as there is hardly a link between industry and agriculture except in isolated

cases like sugar, where productivity has been very effectively linked with the quality of sugar-cane. Much can be achieved by establishing linkages between agriculture and agro-industry and in improving nutrition in an overall way and the status of food technology. Vertical integration of the use of agricultural commodity would contribute substantially in terms of overall progress.

18. One of the most important areas where nutrition and food technology fail is in the utilization of agricultural produce to full potential.
19. Food science and technology could significantly contribute not only in improving the status of agricultural commodities in terms of higher utilization, but more significantly to better nutrition, as many of these commodities could contribute greatly to nutrition. All processes would not necessarily eventually produce items for human beings; there are many innovations and technologies that could use agricultural commodity through cattle feed, fertilizer, and fuel. Some studies have been made to prepare balance sheets of all inputs in agriculture and outputs. Through such studies, one could perhaps evaluate, if not the cost benefit, at least the nutrition-benefit ratio or nutrition budget for important agricultural commodities. If this were done, we could perhaps achieve a proper interphase of agriculture, nutrition, and food science.

The cost of production of most agricultural produce is higher than the purchasing power of the masses. This creates a dilemma.

On the one hand there is a comfortable stock position of over 2 million tons of food grains, yet more than 40 per cent of the population are living below the poverty line and have no means to buy food for a long time to come. In this context, therefore, to consider integrated planning of agriculture and nutrition may seem rather unrealistic. But despite this constraint, anything that could be achieved through interlinkages between policy-planning and management of integrated agriculture could evolve into a pattern that would in some measure serve the nutritional needs of the people.

MISSING LINK

Agriculture, nutrition, and food science have seldom been discussed together. Somehow, discussions on overall planning and policies for agriculture have remained isolated from food science and nutrition. However, taking the overall need as a working base, nutritional inputs and requirements are projected so as to work out per capita availability and consumption. Agricultural production will continue to remain short in relation to overall nutritional requirements.

Advances in food science and technology have been made, and various appropriate technologies applied. It is disheartening to observe that despite the existence of a very large fund of knowledge for bringing about technological changes, overall adoption of new technologies that involve agricultural produce has been slow, even when acceptance of the new technology is considered to be in national interests.

POLICIES AND PLANNING

Is integrated management of agriculture policies and planning feasible in India? Agriculture is a state subject. No amount of policy planning at the central level can be effectively implemented uniformly on an all-India basis. This means, therefore, that policies will have to be broad-based and that implementation will occur according to the specific needs of the area, or, for that matter, the crop system. Co-ordination and integration at an all-India level have been attempted through well-devised interactions between policy-makers and the implementing agencies. Repeatedly over the years policymakers have recognized that planning should be at the base level, and, therefore, policies should be evolved at national level. Under the circumstances how do we go about planning an all-India crop pattern?

For certain categories of production, such as those that are of a perennial nature, some planning is possible since one can estimate the areas under such plantations. But when it comes to general agricultural crops, the pattern of production would vary depending on many factors, primarily decision-making on the part of the individual farmer. Planning in agriculture therefore must establish linkages with the farmer, and some means will have to be found to involve him in the adoption of the plan: otherwise no worthwhile policy can be evolved. It will thus be necessary to know how the farmer will react to various policies in deciding his own plan.

We will also have to determine what his options are, as well as his reactions to any new ideas, and how far he is convinced of the significance of things considered beneficial. So far no attempt has really been made to have the primary producer comprehend any cost-benefit ratios, or to know how he would benefit by adopting a policy strategy evolved at district, state, or central level. In the present socio-economic and political system, it would indeed be important to establish full rapport with the primary producer. This can be achieved to a large extent by interacting first with selected farm leaders, and have these leaders transmit to the farmers the overall need for policies and planning in the total national context. Through such means, there could be widespread adoption of agricultural policies and planning.

However, it is recognized that even with the diverse nature of a subcontinent like India, considerable success has been achieved in improving the overall agricultural situation. The Green Revolution, with its significant adoption of high-yielding varieties all over India, has many lessons for us. Perhaps the techniques employed in extension need much wider appreciation and adoption. Although there may be shortcomings, agricultural development has in itself been epoch-making.

INPUTS

With the high population growth rate in India, there would have to be continuous effort to improve food production. Agricultural production can be considerably increased in India because circumstances are favourable. It has been shown that a very high rate of crop productivity can be achieved, and from the achievement of the research stations one can visualize how much potential there still is to improve overall production. The lab-to-land programme can provide many examples of feasible adoption of high productivity.

The basic need for any human being is food, and in the Indian context, even shelter and clothes are less important. If we decide as a policy to have adequate food for everyone, an all-out attempt must be made to attain the needed production through agriculture by way of a plan that would match agricultural production with nutritional needs. This is feasible, but it does seem that our policy perhaps lacks this recognition of a need for a national nutritional uplift. Some of the policies adopted are discouraging, and do not seem to help the attainment of a higher rate of productivity.

Inputs like seed, irrigation, fertilizer, and pesticides are the basic requirements for attaining a high rate of production. Take for example the use and cost of fertilizer. Since fertilizer is essential to productivity, in the initial stages all encouragement was given to the use of fertilizer. However with the recent considerable escalation in cost, the overall demand for fertilizer has decreased, and this has substantially (10 per cent) lowered agricultural production. Is this in the national interest? would it not have been better to continue to encourage the use of fertilizer and help the farmer achieve higher rates of production? Similarly, for other inputs like pesticides, power, and fuel for irrigation, the policy should be to minimize their costs. It is a tragedy that whenever there is a shortage of power, the first restriction in most states falls on the rural feeders; this restricts use of irrigation, at a time when it is most required.

Certain advanced countries, and even developing countries, have helped achieve higher rates of productivity in agriculture by assuring a supply of inputs at very low cost. Even in India, if we were to produce much more of any foodstuff, it could be an export earner. Considering the overall potential of agricultural development, prospects for India becoming a net exporter of agricultural produce rather than having to depend on imports are very good.

The rate of productivity in agriculture can also be further increased by giving special incentives to those who attain a higher-than-average rate of productivity. Such incentives would attract many farmers, and might give the direction needed to attain specific production aims of any commodity. Many forms of incentive have been designed for industrial development and for export, but in the area of agriculture such forms are very few. In reality there is a negative approach to higher agricultural production, since urbanites and policy-makers continue to scoff at any successes made in agriculture.

AGRICULTURAL PRICES COMMISSION

The Agricultural Prices Commission (APC) is an important agency that can be effectively used as a mechanism of planning and of attaining some integration in agriculture. This role has not so far been fully visualized for the APC. It deals only with working out price support for selected agricultural commodities. To be used as an instrument in policy implementation, APC must broaden its mandate, since it is the prices of agricultural commodities that largely determine the decisions of the farmer. So far the APC has perhaps not given consideration to the nutritional aspect of agricultural commodities. No weightage seems to have been given directly to nutritional need in fixing prices, though in recognition of the overall shortfall in such agricultural commodities as oilseeds or pulses, higher incentive prices have been given. In recommending an overall "food basket" of balanced nutrition for the country, it may not be feasible to advise APC to encourage higher prices for any products that would improve the overall nutritional needs of the nation. As a matter of policy, however, a recommendation should be made that a nutritionist may be included as a member of the APC, with the intention of influencing pricing policies from the point of view of nutrition.

INTEGRATION OF POLICIES

There is a serious lack of integration in the formulation of policies by various ministries. It would appear that often the exigency of the situation is overrated and the perspective plan is sacrificed. To give an example, the policy decision that has been made in regard to sugar price, levy excise duty, and export, in spite of integration at the highest level, suffers from lack of perspective, and has not taken into consideration the high potential for sugar production nor its effect on producers and consumers. Similarly, the decision to import 1.5 million tons of wheat as an expedient reaction to a particular situation may have a long-term effect on the production of wheat, in which India has attained an international record of productivity. would it not be possible to evolve a long-term policy, by which the farmer would be assured of prices that are reasonable, which is linked with any escalation (such linkages are in operation for all dearness allowances) and which is guaranteed for a period of five years? This would enable the farmer to work out a plan of rotation according to his own feasibility.

Another example is the price of milk. Most urban areas in India now have public sector dairies. Milk prices of these dairies are artificially kept low, resulting in substantial losses to the exchequer and, of course, in no significant improvement in the purchase price of milk. There is therefore a lack of any incentive for a higher production of milk, and this would naturally reduce the overall availability of milk and milk products. Not only is the pricing policy affecting milk production, it is also affecting the nutrition of people.

CREDIT

Credit again is an instrument that can really help attain higher agricultural production, but this instrument has not been properly used. General availability of credit is based on the cost of specific crops. However, only a few banks have linked a higher rate of productivity to higher credit. Such linkages between credit and productivity would enable the farmer to use effectively more inputs and improved technology.

Apart from short-term credit for the crop, there is also a real need to organize better credit linkages for land development, irrigation and such post-harvest facilities as storage, handling, and marketing. This has not been recognized. Agricultural production would be diluted through large wastages for lack of post-harvest handling facilities. Damage to perishable commodities would be especially higher without proper post-harvest facilities at the farm level and all through the marketing chain. The processing and marketing of various agricultural commodities deserve better terms of credit. With many of them, the producers' interest can be protected only if commodities are properly preprocessed, packaged, and marketed. However, credit is sometimes restricted and is now given at a lower rate of interest only towards production. But if full benefit of greater production is to be acquired, extension of credit to the later steps of handling, packaging, and transport would not only protect many food commodities from deterioration, not only in terms of quantity but of product quality and nutrition. Much could be done in preventing wastages if easier terms of credit were made available for these purposes. Not only must credit be provided, but incentives may have to be designed to improve the overall handling of all commodities.

MARKETING

Planners often consider agricultural production as a base for all policies. This is highly erroneous. A country that is predominently rural, with an obvious lack of rural development, suffers when the infrastructure needed for proper marketing of agricultural commodities is missing. Post-harvest handling, processing, and marketing of agricultural commodities influence individual nutrition. In other words, it is not sufficient to evaluate the availability of nutritional foodstuffs by considering just the overall production of agricultural commodities. It is necessary to understand how the marketing of food has been organized and how the needs of each consumer have been met.

A study of the public distribution system through ration shops, of the Food Corporation of India and of other agencies that are handling such programmes as Food-for-Work, shows how ineffective are the means to properly protect food and service the consumer. Not only are the costs high, but the deterioration of food in terms of quality and nutrient content is very evident. Surely there is a need for much better understanding of how post-harvest technology could be the means for improving substantially the quality of food.

It may also be relevant to ask whether it is necessary to continue a policy of rationing, recognizing that there is an overall significant reduction in the quality and nutrition of food. Why would it not be possible to replace the entire programme by one of Food-for-Work, which has relevance to the use of manpower and the employment of food as a resource? Extension of the Food-for-Work programme couid also generate employment in the private and public sectors, or enhance a self-employed status. This programme could perhaps adopt nutritional criteria as a means for improving the nutritional status of the lower strata of the community .

Market yards and mandis could also be used in monitoring and evaluating standards and the quality of agricultural commodities. This would also help in improving handling, storage and processing of the trade commodities.

Agro-industries are on the threshold of an expansion programme. However, it is unfortunate that most agriculture commodities needed by these industries are not produced in accordance with the requirements of industry since there is hardly any link between industry and agriculture. An isolated exception is sugar, where productivity has been very effectively linked with the quality of sugar-cane. Establishing linkages between agriculture and agro-industry would help to improve the status of food technology and most likely nutrition. Vertical integration through use of an agricultural commodity would contribute substantially to overall economic progress.

INTEGRATED USE OF AGRICULTURAL COMMODITIES

An important area affecting both nutrition and food technology lies in the utilization of agricultural produce to full potential. In developed countries, producers receive maximum benefit out of such total utilization of agricultural produce. This is achieved through vertical integration, utilizing each and every by-product of agriculture. Food science and technology could significantly contribute to this end of improving the status of agricultural commodities in terms of higher utilization. Moreover, many of these commodities could contribute greatly to nutrition. It is not necessary that all processes produce items for human use; there are many innovative technologies that could use agricultural commodities as cattlefeed, fertilizer, and fuel. Some studies have been made to prepare balance sheets of all inputs into agriculture and its outputs. Through such studies, one can perhaps evaluate, if not the cost-benefit at least the nutrition-benefit ratio for important agricultural commodities. If this were done, we could perhaps achieve a proper interphase of agriculture, nutrition, and food science.

REGIONAL NUTRITION PROGRAMMES

All the governments of the WHO South-East Asia Region have committed themselves to the goal of health for all by the year 2000, and it is now well recognized that health is the result of an interaction of factors at the individual, family, community, and national levels. No single sector or discipline has the capability of acting on the wide range of conditions determining health or on other factors ranging from individual behaviour to overall development policies aimed at reducing the socio-economic inequalities present in every society. From this recognition, the so-called multisectoral actions have resulted, apparently as a panacea to solve all problems. These actions, however, have been difficult to implement, as there are many conceptual, administrative, and political constraints in this type of co-ordinated effort.

Very recently, during a consultation meeting of countries in the region, the food and nutrition strategies of each one were presented and discussed. It was clear that ways and means to approach a solution to the problem were still being viewed with a strong sectoral and professional bias. This has resulted in a myriad of programmes, as many as there are professional disciplines, many of them not fully related to the reality of the problem and its determinants. The agricultural sector has concentrated heavily on food production, with impressive successes in improving the food balance sheets but with limited effect in improving the diets of lower income groups, unable to compete with the richer farmers, for example, in improving productivity (International Centre for Tropical Agriculture 1974). Also, there have been many experiments on developing proper foods through food technology research, but very few of these formulae have had any impact on the dietary intake of low income groups except in developed countries (Behar et al. 1971). Within the health sector, there have been several types of intervention, including education, rehabilitation, and food distribution, without regard for the fact that those most in need are not usually reached and do not take part in these programmes (Ghassemi 1980). In the area of education, many attempts have been made, with significant results in getting information to the populations but with very limited effect on changing behaviour.

Many of these programmes have failed through lack of proper implementation, and many by disregarding the intrafamily, interfamily, and interregional processes normally taking place for the benefit of some and to the detriment of others, or as a social response for survival. Indeed, individual and social behaviour is a product of adaptation for survival and not necessarily that of ignorance and backwardness (Nabrarro 1982). Apparently, the intersectoral approach requires that we focus on the problem to be solved, understand its various aspects, and only then define the role of each sector or discipline. As the processes may be different for different population groups, in differing ecological regions, and under different political circumstances, a different set of sectors may be called upon to help solve the same problem in different locations.

OVERVIEW OF THE REGIONAL PROBLEM AND ASSOCIATED FACTORS

Countries in the WHO South-East Asia Region (Bangladesh, Bhutan, Burma, India, Indonesia, the Democratic People's Republic of Korea (DPRK), the Maldives, Mongolia, Nepal, Sri Lanka, and Thailand) contain roughly 25 per cent of the world's population, most of the ethnic groups, a large representation of religions and languages, several types of governments, and a full range of ecological systems. These variations occur not only from country to country but within countries, making any mean average almost worthless. However, it is possible to establish qualitatively some comparisons of the magnitude of the problem of malnutrition as well as of some of the conditioning factors.

The growth process of children from conception to puberty has been an accepted indicator of nutritional status. It must be recognized that it is not a specific indicator of nutrient intake although it may be specific for tissue nutritional status and, as such, may reflect the adequacy of nutrient intake, or nutrient absorption and/or nutrient uptake. Furthermore, growth is of use only as it reflects the individual development and function as well as that of the family and the community in which the individual lives.

As mentioned previously, there is also a significant variation within countries. The same criteria, when available, demonstrate extreme variations in the magnitude of the prevalence of malnutrition within country boundaries (see FIG. 2 Magnitude of Malnutrition within Asian Country Boundaries (Based on the Mortality Rates of Children Aged One to Four)). Some factors that have been identified as associated with malnutrition (growth deficit) such as nutrient availability, nutrient consumption, income, education, and morbidity, were also collected and analysed in relation to their contribution to the problem. For this analysis, the available unconsolidated data were utilized to determine the strength of the association with nutritional status. Though no strong association was expected, on the assumption that the average for an area or a country represents a large number of universes, each of which has a different set of conditions determining the effect of a given variable. The strongest association with nutritional status appears to be literacy and income.

A more detailed analysis of one of the factors, energy intake, was undertaken using published data supplemented by personal communications. The mean value and standard deviation for energy intake were derived for Bangladesh (Mostafa 1979, Institute of Nutrition and Food Science 1976), India (NNMB 1976-1980; Panikar 1980), Indonesia (Djumadias 1979; Goan-Hong Lie 1976), Sri Lanka (MRI 1976), Thailand (Nutrition Division 1981), the Philippines (FNRI 1980), Colombia (Fundación para la Educación Superior 1976), and the United States (PHS 1979).

Mean Value and One Standard Deviation for Different Developed and Developing Countries or Areas) shows in graphic form the lack of significant differences between energy intake in the United States and that in developing countries or areas within countries. This finding might not have any significance if other similar contradictory observations were not available. The state of Kerala in India has had the lowest per capita intake of energy and proteins in the country and, on the other hand, the best growth performance and infant survival rates (Panikar 1980). A field experience, in Candelaria (Colombia), shows a decline of nutrient consumption with a simultaneous improvement of growth performance and child survival (Pradilla et al. 1975, unpublished). Also the area where nutrient consumption is the lowest in Thailand has the best nutritional status. In the United Kingdom, Nelson (1980) has also reported unexpected differences in growth with an equal intake of energy and proteins. From these observations, one can draw several general and specific conclusions of regional interest:

1. Recommended dietary allowances in practice seem to set levels of intake that cannot be reached by the average population even in highly developed countries. These recommendations should be interpreted in the spirit in which they were intended, recognizing the fact that there is a significant variation in individual requirements.
2. The problem of malnutrition as defined by growth performance represents different combinations of factors . The effects of a given factor maybe minimized or enhanced depending on the other variables present in a specific location. There cannot be a single universal approach, nor is there any vertical or individual approach that will make the same impact in every location or among every population group.
3. Whereas a positive impact can be produced in a "sick" child by supplying him with energy- or protein-dense foods, the preventive measure of increasing food availability at the family level without regard for other factors may have no effect unless the lack of food is the main problem in that specific instance, for example, in the case of famine.
4. With a given dietary intake, growth velocity, including that in the intra-uterine period of life, or weight gain is determined by environmental factors.

5. "Nutrition" programmes can be evolved either after a critical evaluation of ongoing programmes to determine whether they relate to the basic causal factors, or after an analysis of the prevailing conditions and associated factors. Once the problem and the factors are defined, the role to be played by the different sectors can be determined.

RESPONSIBILITIES OF THE HEALTH SECTOR IN NUTRITION

The WHO programme for the South-East Asia Region includes support within the health system of those activities known to have an impact on the growth of children, e.g. their nutritional status, from conception, and of the mechanisms necessary to determine necessary action by other sectors. It is recognized that every country is developing its own programme, that many of the activities may already be under way, or that the interest in them is not as yet sufficient for them to have been included in national priorities or policies. For this reason, the programme is directed essentially toward promoting/strengthening/supporting appropriate mechanisms within the countries to facilitate a multisectoral analysis of the problem and its determining factors, and the development, testing, and evaluation of solutions in line with each country's socio-political reality, to help in attaining the following objectives:

TABLE. Intra-uterine Malnutrition

Problem	Indicator	Associated factors	Determining conditions
Intra-uterine malnutrition	Low birth weight for gestational age intervals	1. *Preconception*	
		Mother height age, pregnancy	Her own childhood: growth and development.
			Reproductive behaviour
		2. *Pregnancy*	
		Maternal nutritional status	Energy expenditure
			Infections or other diseases
			Nutrient intake
		Placental status	Genetic, metabolic
			Infections or other diseases
		Foetal factors	Genetic, metabolic Infections

1. To *decrease the incidence of intra-uterine malnutrition* (as measured by low birth-weight for gestational age). The indicator, along with some recognized causal factors

and determinants, is shown in table 2. The role of each factor and its importance may be different in each specific local circumstance. Whereas food intake, for example, might have critical importance in one location, reproductive behaviour or illness might be more important in another.

2. *To minimize the metabolic, symptomatic, and cultural factors associated with increased requirements and interfering with food intake during disease and convalescence.* Table 3 includes the indicator as well as the causal factors and their determinants. Most of the factors are related to the incidence and duration/severity of disease and especially of those diseases having a significant effect on the growth of children (diarrhoeal diseases, measles, whooping cough, hookworm infestation, etc.). Included also are those factors related to dietary management during illness and convalescence.
3. *To optimize the utilization of locally available staple foods and to collaborate with efforts to increase the food consumption by all members of family.* This includes, in table 4, factors that are related to food habits, infant- and child-rearing practices, health-positive behaviour and negative health practices. The responsibilities here are mainly educational.
4. *To detect early deviations from normal growth, to determine its causes and to take appropriate measures to correct them.* This objective is a supportive one. Table 5 shows that it includes the development of appropriate mechanisms for surveillance of the individual and of communities, the changes of conditions in the country, and the factors responsible for them, so as to be able to take appropriate local, regional, and national actions.
5. *To apply available methodology in the correction of specific nutrition deficiencies.* Among these are goitre, iron-deficiency anaemias, and vitamin A deficiency.

It is clear that most, if not ail, of the activities require a synchronized, concerted effort by different agencies acting in close co-operation to respond to community needs. Most of the hardware, if the term can be used, and sector-specific software are to be supplied by sectors other than health. It is also apparent that unless a network is developed that will include all sectors of society, very few of these activities can be accomplished. This is one of the basic reasons why the development of health services and primary health care, together with community participation, is one of the solid pillars of the strategies that have been agreed upon by governments in the world Health Assembly and later at the sessions of the WHO regional committees held in different regions throughout the world.

MALTED VERSUS ROASTED WEANING MIXES

The search is on for low-cost weaning and young toddler foods that would combine at least some of the desired characteristics of high nutrient density, low bulk properties, utilization of low-cost and widely used cereals, pulses and oilseeds, and traditional processing methods that have the potential of being easily adopted at the home or village level (Desikachar 1980; Chandrashekhara 1980; Narasinga Rao 1977; Gopaldas et al. 1974; Devadas et al. 1974; Reddy 1973; Swaminathan et al. 1973; Swaminathan et al. 1972; Rajalakshmi 1970; Korula et al. 1961).

There is recent evidence (Desikachar 1980) to show that malted ready-to-eat mixes have an overriding advantage over processing methods such as flaking, puffing, parboiling, or roasting in that malting greatly reduces the viscosity of the product thereby leading to increased nutrient or calorie-dense food per unit volume ingested. Young toddlers who have a severe bulk intake constraint are able to consume relatively more of a malted food product at a sitting. An added advantage of malting is that starch is partially pre-digested to dextrins which would help a weanling infant to adjust more easily from a lactose-based milk diet to a starch-based cereal diet.

Very little work is cited in the literature on enhancement of nutrients or reduction in anti-nutritional factors as a result of the final malting process. However, several workers have reported the beneficial effects of germination (intermediate step in malting) in cereals and/or legumes such as increase in thiamin, riboflavin, and niacin levels (Burkholder 1943); folic acid (Babu 1976); ascorbic acid (Patwardhan 1962); iron (NIN 1978); amylase and diastase activity (Jaya and Venkataraman 1980); protease activity (Jain and Date 1975); digestibility (Subbulakshmi et al. 1976); protein efficiency ratio and biological value (Chandrashekhar and Jayalakshmi 1978). Besides decreases have been recorded in phytate (Mandal et al. 1972) and in trypsin activity of haemagglutinin (Subbulakshmi et al. 1976).

Roasting, which is a simpler and more commonly used household and village-level technology, pre-cooks the ingredients used in cereal-legume-oilseed multimixes and increases shelf life and acceptability (Gopaldas et al. 1975). Siegel and Fawcett (1976) have reported that it improves the flavour, texture, and nutritive value of the grains. Studies conducted by Liener (1973) report that most anti-nutritional or toxic effects of legumes such as trypsin inhibitor, haemagglutinin, goitrogenic agents, cyanogenic glucosides, alkaloids, and saponin were eliminated partially or wholly on roasting.

The present paper attempts to give the highlights of a series of studies, undertaken in the Foods and Nutrition Department of the Maharaja Sayajrao University which have helped to shed more light on comparative aspects of roasted and malted foods with respect to their formulation, ease of processing, cost, proximate composition, selected physicochemical characteristics that have a bearing on nutrient intake capacity, storage trials, human acceptability trials, and growth trials. A great deal more needs to be done and many gaps in our own series of experiments need to be filled and knowledge updated.

Since there was a distinct regional preference for the staple wheat (Triticum aestivum), Bengal gram (chick-pea, Cicer arietinum), and groundnut (Arachis hypogaea), most of our work pertains to formulations made from these three food commodities. However, we did work on ragi (Eleusine coracana), green gram (Phaseolus aureus Roxb), and groundnut formulations also, but our human acceptability trials in Baroda clearly showed that ragi-based formulations were quite unacceptable to adult and child subjects.

Pertinent data from four M. Sc. dissertations, (Inamdar 1980; Fotedar 1981; Tajuddin 1981; Master 1981) have been brought together in this paper to throw more light on the positive and negative points of malted and roasted mixes having identical formulations. Individual papers on aspects discussed in this paper are in the process of being published.

FORMULATIONS AND STANDARDIZATION

Formulations incorporated low cost and widely used cereals, pulses, and oilseeds in Gujarat. The ragigreen gram formulation with and without oilseeds was also included as the pioneering work of Malleshi and Desikachar (1979) indicated its superior nutritive value in weaning foods.

The proportions of cereal to pulse or of cereal to pulse to oilseed (v/v) were kept simple, as the ultimate purpose was to transfer the technology to the household and village level. Formulations with a fairly wide cereal-to-pulse ratio of 8:1 were also included as (a) this is roughly the ratio of staple to legumes consumed in the low-income groups, and (b) it has been cited in the literature that the growth-promoting effect of cereal and pulse in the ratio of 4:1 or 8:1 was not found to be markedly different for the weanling rat (Rajalakshmi and Nalini 1966). If this could be substantiated in the human child it would help to reduce the formulation cost. The inclusion of oilseeds was to have a more nutrient-dense formulation; their omission was to reduce cost.

TABLE. Weaning Formulations Studied

	Proportions (v/v)			
	4:1	4:1:1	8:1	8:1:1
Wheat-Bengal gram (-groundnut)				
Malted	v	v	v	v
Roasted	v	v	v	v
Ragi-green gram (-groundnut)				
Malted	v	v	-	-
Roasted	v	v	-	-

Malted Multimixes

The main steps in malting are steeping, germination, roasting and milling. Each step was standardized on the basis of organoleptic acceptance, particularly for taste, by a trained panel of six members.

1. *Steeping*. All seeds gave best results when soaked in an equal volume of water at room temperature (25° - 30° C) for 12 hours. The relative humidity was 71 + 12 %. Our steeping conditions are in line with those reported by Patel (1957) who studied the most practical household methods for steeping and germination of the common cereals and pulses consumed in Gujarat.
2. *Germination*. Wrapping the seeds in damp muslin cloth for germination at room temperature, and arresting germination at 24 hours in the case of groundnut and 48 hours in the case of wheat and Bengal gram was found to be the most acceptable organoleptically .
3. *Roasting and milling*. Roasting was done in an electric oven at 110°C ± 3°C. Groundnuts required 8-10 minutes, wheat and ragi 10-15 minutes, and Bengal gram and green gram 20-25 minutes. The entire malted grains (seed + vegetative part) was milled in the proportions stated in a Kissan Grindmaster (up to 1 kg quantity) or in a regular electric mill for larger quantities. The milled flour was sieved four or five times through a 50 mesh BSS sieve. The loss in residue was about 15 per cent for wheat and ragi, 15 per cent for Bengal gram, or green gram, and about 10 per cent for groundnut. In the case of simple roasted mixtures, only the roasting process followed by milling and sieving was utilized.

Production Problems of Macro Quantities

Producing enough malted or roasted supplements (about 125 9 per serving) for 30 children for seven days at a time in the laboratory presented problems of non-availability of suitable utensils and space, and highlighted the need to fabricate simple trays for uniform steeping and germination, and simple temperature-controlled roasters to produce approximately 200 kg of mix per batch. In a typical village situation 90 kg of multimix per week would be required to feed 100 infants/ toddlers per averaged-sized Indian village (total population of 1,000).

Proximate Composition

The proximate composition of the most nutrient-rich wheat-Bengal gram-groundnut (WBG) and ragigreen gram-groundnut (RGG) formulations (4: 1:1). Heat processing either by malting or roasting did result in appreciable loss in ash and fat as compared to the raw mixes. However, energy and protein levels were well within the accepted range of 12-14 9 for protein and 350 kcal or more per 100 9 of formulation for supplementary feeding.

TABLE. Proximate Composition of Malted and Roasted WBG and RGG Formulation[a]

Formulation (g/100 g)	Moisture	Ash	Fibre	Fat	Protein	Carbohydrate
WBG -4 :1:1b						
Malted	18.36	0.69	2.27	7.34	14.35	56.99
	(24)	(- 44)	(43)	(- 30)	(5)	(- 2)
Roasted	17.54	0.57	1.98	8.27	14.24	57.4
	(18)	(- 54)	(24)	(- 21)	(4)	(- 1)
RGG -4 :1:1b						
Malted	16.67	1.75	2.52	8.67	13.27	57.12
	(20)	(- 42)	(- 3)	(- 9)	(7)	(- 2)
Roasted	17.78	2.14	2.89	7.97	12.56	56.66
	(28)	(- 29)	(12)	(- 16)	(1)	(- 3)

a. WBG = wheat, Bengal gram, and groundout; RGG = ragi, green gram, and groundnut.

b. Figures in parentheses show percentage changes from the analysed values of raw mix.

VISCOSITY MEASUREMENTS OF A 10 PER CENT COLD-PASTE SLURRY SOLUTION OF VARIOUS WBG FORMULATIONS

Viscosities of a 10 per cent cold-paste slurry of the various malted and roasted formulations of WBG determined in a Brookfield's viscometer (LVF model no. 50). The most striking feature of this experiment was the significantly lowered viscosity readings of malted formulations vis-à-vis their roasted counterparts. Our findings support those of Malleshi and Desikachar (1979) reported for malted versus unmalted ragi:

green gram (70:3û) in a 10 per cent hot-paste slurry solution. The decrease in viscosity of our malted formulations with only a slight increase in volume as compared to simple roasted counterparts, clearly suggests that malted formulations with low bulk and high caloric density can be more easily consumed by toddlers than their roasted counterpart formulations. As described later, this was borne out also by acceptability and intake trials among Baroda balwadi (children's centre) children.

TABLE. Viscosity Measurements of a 10 Per Cent Slurry Solution of WBG[a] Malted and Roasted Formulation[b]

		Volume gm/ml	Viscosity Cp units
Form of mix	**Ratio**	**Slurry 10 %**	**Slurry conc. 10 %**
Malted	4:1:1	109	5
Roasted	4:1:1	106	9
Malted	8:1:1	110	7
Roasted	8:1:1	108	12
Malted	4:1:0	112	8
Roasted	4:1:0	107	14
Malted	8:1:0	110	10
Roasted	8:1:0	108	15

a. WBG = wheat, Bengal gram, and groundnut.

b. Measured in a Brookfield's Viscometer (LVF model no, 50).

STORAGE TRIALS

Since the WBG formulation was most relevant to Gujarat, the keeping quality of the most nutrient rich formulations of WBG (4:1 :1) was studied over 28 days' duration at room temperature. One hundred grams of each ready-to-eat mix (malted or roasted) were packed and sealed in polyethylene packets of 200 gauge (7 cm x 10 cm). Packets were drawn for analysis at 0, 7, 14, 21, and 28 days. The samples were analysed for moisture (AOAC 1960), reducing sugar (Hawk 1965), alcoholic acidity (ISI 1957), peroxide value (AOAC 1975), and bacterial contamination (Mackie and McCartney 1956). The maximum limit for moisture content irrespective of season has been laid down as 14 per cent (PFA 1977). The malted WBG was acceptable up to the fourteenth day; the roasted was well within specifications of the Indian Standards Institute (ISI) throughout the trial. The free sugar level was about three times as high for the malted WBG as for the roasted on the twenty-eighth day. There are no ISI standards for acceptable shelf-life. However, the saccharifying effect of malting could be considered a plus rather than a negative point for a weaning mixture. The alcoholic acidity level of the malted WBG was more than twice that of the roasted on the twenty-eighth day.

TABLE. Storage Trials for Different Malted and Roasted Formulations of WBG (4 :1:1)

Storage period (days)		Moisture (%)	Free sugar (%)	Alcoholic acidity (%)	Peroxide value m moles per kg fat	Bacterio logical examination
0	malted	10.44	12.01	0.083	3.0	16,000
	roasted	7.08	2.18	0.043	2.1	1,250
7	malted	12.14	12.24	0.083	7.2	20,500
	roasted	7 52	2.74	0.048	3.8	28,000
14	malted	13.98	12.63	0.093	11.1	41,500
	roasted	7 62	3.13	0.055	4.5	36,000
21	malted	14.65	13.08	0.121	14.8	*
	roasted	8.10	3.67	0.060	5.2	43,000
28	malted	14.78	13.12	0.141	183	*
	roasted	8.38	4.11	0.063	5.8	53,000
PFA	values	14.00	**	0.120	10.0	50,000***
				0.100***		

* Innumerable colonies.

*' No standards

*** 151 standard.

The specification of the Indian Prevention of Food Adulteration Act (PFA) as well as of the Indian Standards Institution is 0.10 to 0.12 per cent. Hence, the malted WBG was acceptable until about the twenty-first day, while the roasted WBG was well within ISI/PFA limits. The peroxide value for M-WBG was over three times that of R-WBG on the twenty-eighth day. The PFA standard is 10 millimoles/kg fat. According to this stipulation M-WBG was acceptable only for seven days, while R-WBG was well within the limits even on the twenty-eighth day. The bacteriological plate count for malted was within ISI specification (1973) of 50,000 for 14 days; it was 14 days for the roasted. It was, therefore, concluded that the malted had essentially a shelf-life of seven days and roasted of 28 days. Our data (not presented here) indicates that when oilseeds were omitted from the malted WBG (4: 1 :0) formulation, the shelf-life could be stretched to 14 days.

COST OF FORMULATIONS

The cost of the WBG (4:1:1) formulation per kilogram of ready-to-eat mix in 1979 was Rs 3.20 for the malted and Rs 3.00 for the roasted. It was slightly higher in 1980; Rs 3.40 per kg for malted and Rs 3.20 per kg for the roasted. Corresponding costs for the RGG (4:1 :1) in 1980 were Rs-4.17 for the malted and Rs-3.76 per kg for the roasted. Even if one doubles the ingredient cost to include overhead, packaging, logistical, and other related costs, the M-WBG or M-RGG would still be far below the cost of commercial formulations such as Farex (Rs 40 per kg) or Bal-Amul (Rs 34 per kg) or Nestum (Rs 16 per kg). A product test conducted on WBG (4:1:1) among railway worker families indicated that they were prepared to pay as much as Rs 4.00 per kg, or 50 paise per 125 9 packet.

ACCEPTABILITY TRIALS IN BARODA

Acceptability was evaluated by (1) organoleptic trials among adults (Amerine et al. 1965), and (2) feeding acceptability trials among balwadi children (Grewal et al. 1974).

1. Organoleptic or sensory evaluation. The Triangle Test followed by the Score Card Technique were used. The first test was to select a discriminating panel who could detect two identical samples out of a triangle of three samples. Ten college students were thus selected for the Score Card Test. The formulations were scored on a seven-point hedonic scale. For minimum acceptance, a formulation had to obtain a minimum of 60 per cent of the maximum score in each characteristic of taste, appearance, texture, aroma, and after-taste, as well as on the overall score. The test was conducted thrice to arrive at consistent results. The results are presented in table 5. It can be seen that there was no significant difference in acceptability between the malted and roasted formulations for WBG (4: 1:1). However, of the malted and roasted formulations for WB (8:1), the malted was significantly preferred to the roasted for taste and aroma; the roasted was significantly preferred to the malted for texture.

2. Acceptability trials among balwadi children (three to four years). Acceptability was evaluated by three criteria: minimum intake of RTE per sitting; request for second helpings; and extent of plate waste and/or spillage.

TABLE. Acceptability Trials in Baroda on Formulations of WBG by Balwadi Children (three-four years)[a]

Ratio	Mean consumption (gm)		Paired "t" value
	Malted RTEs	Simple roasted RTEs	
4:1:1	83.7 ± 2.49	67.9 ± 1.23	5.95[b]
8:1:1	81.2 ± 4.73	62.8 ± 1.33	9.01[b]
4:1:0	75.6 ± 2.44	59.2 ± 1.49	7.24
8:1:0	75.6 ± 2.00	56.8 ± 1.95	8.64[b]

a. Values are mean + SE. b. P = .05.

Twenty-seven balwadi children participated in this 24-day trial with each formulation being tested for three days consecutively in random order. Each child was offered an initial serving of 100 9 formulation of WBG (either 4:1:1 or 8: 1:1) or WB (4: 1 or 8:1). Each formulation was sweetened with 25 per cent by weight of jaggery. Each child was offered an initial serving of 100 9 formulation. A second helping of 25 9 was served if requested. Records were maintained per child subject. The results of mean intake of malted and roasted of each formulation are represented in table 6. It is clearly seen that for any formulation, the paired " t" test established that malted formulations were significantly preferred to roasted formulations. The maximum mean intake was on M-WBG (4: 1:1) at 83.7 2.49 9 and the least was on WBG (8: 1 :0) at 56.8 1.95 ფ Non-incorporation of oilseed significantly reduced intake, most probably due to lessened palatability.

ACCEPTABILITY TRIALS IN PONDICHERRY

Preliminary trials conducted on foundling home and low-income group children (two to five years) in Baroda established that RGG formulations were rejected. Consequently, an inter-comparison of acceptability between malted and roasted formulations of WBG and RGG could not be carried out at Baroda. RGG formulations, on the other hand, were found to be highly acceptable in Pondicherry, where ragi is the staple.

This clearly highlighted the regional preference for specific cereals and pulses. An acceptability feeding trial was planned among 18 dewormed balwadi children (two to five years of age) for 14 days. Children were randomly assigned into two groups. Both the malted and roasted RGG (4:1:1) were tested by all child subjects for seven days each. The criteria developed for acceptability were:

1. The child should consume at least 75 9 out of a 100 9 serving of formulation, at one sitting, in one hour.
2. The formulation should not cause any ill-effects like diarrhoea or vomiting.
3. This amount should be consumed in addition to their home diet.
4. Attendance should be at least 75 per cent of the total period of testing.

The set of criteria were satisfied for both roasted and malted mixes. It was clearly established that acceptability (as assessed by mean intake) was significantly more for malted RGG.

TABLE 7. Acceptability Trials in Pondicherry on Formulations of RGG on Balwadi Children (two-five years)

Type of mix	Mean daily consumption (gm)	S.E.	"t" value[a]
Malted RTE mix	120.6	8.68	2.1594[b]
Roasted RTE mix	102.4	4.71	

a. Paired "t" test

b. Significant (P = .05).

RAT GROWTH AND PROTEIN QUALITY TRIALS

Eight formulations, four of malted and roasted WBG (4:1:0, 4: 1:1) and four malted and roasted RGG (4:1:0, 4:1:1) were evaluated against a 10 per cent casein control diet. Seventy-two weanling albino rats of Wistar strain were divided into nine groups of eight each. The parameters to evaluate growth and protein quality were (a) food efficiency ratio (FER) defined as amount of formulation required to bring about 1 9 weight increase in the animal in 28 days; (b) the protein efficiency ratio defined as the weight gain per 9 protein casein; (c) total serum protein and albumin/globulin ratio; (d) liver protein; (e) haemoglobin levels. At a glance it is seen that the protein quality of casein cannot be matched by either malted or roasted formulations of WBG or RGG. However, between malted and roasted formulations, the malted formulations score much higher than the roasted counterparts, for each and every parameter studied. Within malted RGG and malted WBG, malted RGG was found to be superior. Within the malted WBG formulations, WB (4:1) was actually found to be superior to WBG (4:1:1), thereby indicating that incorporation of oilseed did not exert any beneficial effect with respect to growth or improvement in the protein status of the rat.

TABLE. Evaluation of Growth Promoting and Protein Quality of Malted[a] and Roasted[b] Multimixes on Rats[c]

	FER[d]	PER[e]	Total serum protein g/dl	A/G Ratio	Liver protein 9/100 g	Haemoglobin g/dl
Casein	2.9	3.38 ± 0.190	7.34 ± 0.293	1.42 ± 0.084	12.29 ± 0.194	12.15 ± 0.495
R-WB	15.9	0.63 ± 0.137	5.55 ± 0.183	0.86 ± 0.078	6.05 ± 0.262	8.57 ± 0.475
M-WB	5.0	2.02 ± 0.183	6.21 ± 0.214	1.59 ± 0.082	7.52 ± 0.270	11.12 0.615
R-WBG	19.2	0.53 ± 0.102	5.52 ± 0.114	1.46 ± 0.101	6.58 ± 0.217	9.49 ± 0.507
M-WBG	6.0	1.64 ± 0.181	5.50 ± 0.196	1.16 ± 0.065	8.23 ± 0.246	9.66 ± 0.229
R-RG	4.6	2.14 ± 0.180	5.77 ± 0.513	1.29 ± 0.079	8.51 ± 0.150	10.02 ± 0.446
M-RG	4.5	2.20 ± 0.113	6.73 ± 0.462	1.55 ± 0.107	9.53 ± 0.262	10.24 ± 0.276
R-RGG	4.7	2.03 ± 0.341	6.63 ± 0.192	1.52 ± 0.073	10.06 ± 0.414	10.08 ± 0.390
M-RGG	4.7	2.07 ± 0.130	6.91 ± 0.196	1.63 ± 0.110	11.19 ± 0.220	10.79 ± 0.227

a. M = Malted.

b. R = Roasted.

c. Values are mean ± S.E.

d. Food efficiency ratio.

e. Protein efficiency ratio.

PROSPECTS AND PROBLEMS

Our studies, thus far, have established the superiority of malted cereal-pulse-oilseed formulations, whether they be WBG or RGG over their roasted counterparts. However, regional preferences for staples and pulses will have a very definite influence on the future development of malted multimixes. The proximate composition of the most nutrient rich formulations of malted or roasted WBG or RGG (4:1:1) or even those that omit oilseed, reveal that they are within the acceptable limits of 350 kcal and 10-12 g protein per 100 g of food supplement. Acceptability trials among young balwadi children have further established the significant superiority of the regionally preferred malted formulations over their roasted counterparts. Per sitting intake levels play a crucial part in bridging if not completely filling recommended daily allowance requirements of the infant and the young toddler and hence any village-level technology that supports this important aspect needs to be favoured.

The ingredient cost for either malted or roasted could be considered to be reasonable at about Rs 3.00-Rs 4.17 per kg in 1980. Even if one doubles this cost for overheads, packaging, logistics, and other related costs, it would still work out to be one of the most economical weaning mixes in the present day Indian market.

Our rat growth and protein quality trials on malted and roasted formulations of WBG and RGG again favour the malted formulations whether they be with or without oilseeds. Although long-term growth trials have yet to be conducted on WBG (4:1:1), some indication of the benefits of malted mixes for the very young child are available from earlier studies conducted in the department. Gujral (1968) conducted a eight month growth trial on Baroda children one to three years, wherein a simple malted gruel of wheat and green gram (1:1 by weight) was offered at the level of 60 g per day. In addition, a spoonful of vitamin A enriched oil at the level of 3 g per day was offered per child. The results indicated improvements in the anthropometric, clinical, biochemical, and radiological status of the conjee-supplemented children. However, no comparative growth data on a roasted conjee of similar formulation was available.

In sum, the pioneering work of Korula et al. (1961), followed by the recent in-depth studies of Desikachar (1980), the infant and young toddler feeding trials of Swaminathan (1980), and our own clearly indicate the tremendous potential of malted cereal: pulse: oilseed multimixes in the Indian context; particularly as germination of cereals and legume grains is a universal and traditional practice in our land.

Although the prospects for malted multimixes are bright, yet, there are two major stumbling blocks which must be overcome before one can consider the malting process as an appropriate village-level technology.

The first concerns its poor shelf-life; especially so when oilseeds are incorporated. The second concerns the rather lengthy and laborious steps in processing, namely, steeping, germination, roasting, and milling. An average village home is not likely to have the space (for drying the germinated cereals, pulses, and oilseeds) or the stensils. Even simple roasted multimixes were not made by the majority of Madhya Pradesh rural mothers (Gopaldas et al. 1975).

Hence, even if one overcomes the keepability problem by omitting oilseeds in the formulation, yet, one will have to decide where? how? in what manner? by whom? and at what cost? these malted multimixes can be produced on a scale to feed at least 100 (0-3 year-old children) day-in and day-out in an average Indian village (total population of 1,000). Common sense suggests that it would have to be a centrally operated village, block, or district facility. There is, therefore, an urgent need to develop simple inexpensive equipment for steeping, germination, roasting, and milling for at least a village-level operation.

In conclusion, we see more prospects than problems with malted multimixes. Studies are planned or are in progress in the department to evaluate the acceptability of WBG and WB formulations with sesame seed among geriatrics; to determine its rehabilitative properties in the local Nutrition Rehabilitation Unit of SSG Hospital, Baroda; and to study its enhanced nutritional properties, if any, on fermentation, and with and without the incorporation of green leafy vegetables.

COUNTRY REPORTS

The land area of Sri Lanka is about 16.2 million acres (25,000 square miles) of which 5.5 million acres are used for permanent agriculture, 9.3 million acres are under forest and 2 5 million acres are under chena, or shifting cultivation. Agriculture is of prime importance in the economy of the island, accounting for about a third of the gross national product. Over 80 per cent of the export earnings depend on the export of tea, rubber, and coconut products.

The growth rate of the agriculture sector showed a decline from 5.4 per cent in 1978 to 2 per cent in 1979, but recovered to 3 1 per cent in 1980.

This drop is attributed to adverse weather conditions, which caused a significant setback in overall production, mainly of tea, rubber, and coconut. In 1981 the growth rate in the agriculture sector is expected to further rise due to increases in production of paddy and some other field crops.

Basically, the agriculture in Sri Lanka can be resolved into a domestic agriculture sector, and a plantation or three-crop sector consisting of tea, rubber, and coconut plantations. Broadly, the agricultural activities in Sri Lanka are governed by two major seasons, the maha and yala. The maha is characterized by the north-east monsoon and extends from October to February, while the yala extends from May to September, when the southwest monsoon is active.

The wet zone, comprising the central hills and the south-western part of the island, and the dry zone, comprising the northern and north-eastern part of the island, constitute the major agro-climatic zones of Sri Lanka.

It is possible to identify 21 agro-ecological regions, or niches, within this major classification. The wet zone is intensively exploited, with almost 70 per cent of the land area under permanent agriculture. A major part of the dry zone is now being developed.

Most of the land is cultivated in the ma ha season under rain-fed conditions , with supplementary irrigation. During yala, cultivation is predominantly under irrigation. The total investment envisaged on the Mahaweli River-diversion project, the longest irrigation project in Sri Lanka, is Rs 28,000 million (1980-1985) representing almost 38 per cent of the total capital budget of the public investment programme.

The total investment in the irrigation subsector, apart from the Mahaweli Development Programme, is estimated at Rs

3,566 million. It is expected to bring 50,000 acres (20,250 ha) under irrigation by 1985 and improve irrigation facilities over another 200,000 acres (81,000 ha) of existing irrigable lands. Besides, the contribution made by small ground-water irrigation projects under the District Development Rural Integrated Programme is around Rs 1,400 million for 1981-1985, benefiting about 115,000 acres (46,575 ha). In this context it is clear that the Government is according the highest priority to agriculture infrastructure development.

The principal field crops grown in Sri Lanka are rice (paddy), coarse grains (maize, sorghum, and millet), pulses (cow-pea, green gram, soya, groundnut, black gram), starch and tubers (potatoes, sweet potatoes, manioc), agro-industrial crops, mainly sugar and oilseed crops (gingelly, kapok, etc.), and minor export crops (cinnamon, cardamon, nutmeg, and cloves).

The land area cultivated under paddy is 597,624 ha with an estimated production figure of 80.8 million bushels. Average yield per acre is approximately 56 bushels/acre. The area cultivated with subsidiary food crops during maha is about 126,331 ha while in Yala due to reduced rainfall this shrinks to about 42,110 ha. Over 2,000 ha are planted to sugar-cane and about 4,750 ha to minor export crops. The cropping intensity of paddy is 175 per cent in the wet zone and 110 per cent in the dry zone.

Thus while working towards the goal of achieving self-sufficiency in rice production it becomes evident that emphasis has to be given to increase the land acreages and cropping intensities of the other field crops, specifically the cheap grain, root, pulse and tuber crops (CGRPT). The lowest deciles of the income group, though dependent heavily for their calories on rice and coconuts, will have to depend to a greater extent than before on cheaper proteins and cereal sources in view of the rising prices of basic food commodities.

While liberalization of imports, reduced tariffs and duties, and increased food aid coupled with better agri-support services have resulted in a greater availability of food items, it is extremely important to safeguard the local producers against ad hoc imports of food items that could be counter-productive to the food production effort, since in the long term it is the low cost food production sector that will provide the answer to the poorer income segments of the population.

CALORIE AVAILABILITY AND FOOD CONSUMPTION

Food consumption in terms of per capita availability is given in table below. The pattern of food consumption has remained quite stable over the last three decades. The major sources of energy derived continues to be from plants and to a considerably lesser extent from foods of fish and animal origin.

TABLE. Per Capita Availability of Energy-yielding Foods

	Ten-year averages in kg/per capita			
	1950s	**1960s**	**1970s**	**Most recent year-1979**
Rice	92.9	100.6	96.8	91 9
% total calories	48.6	47.6	41.5	37.9
Wheat	24.7	25.2	36.8	37.9
% calories	15.4	16.6	15.3	15 6
Maize	7.5	1.1	1 4	1 3
Millet	2.3	1.8	1.2	0 9
Others	0.7	0.3	0.2	0 1
Total cereals	128 2	129.0	135.4	132.1
% calories	50.8	60.8	58.0	544
Yams and potatoes	26.4	30.0	38.0	35.2
Nuts, oils, and fats				
% calories	11.8	16.0	20.5	22 7
Sugar				
% calories	8.7	9.6	6.0	7.5
Total kcal/capita/day	2,120.0	2,127.0	2,235.0	2,317.0
% energy from vegetable sources (6 + 7 + 8 + 9)	84.8	86.6	91.3	90.7

Sources: Department of Census and Statistics; De Mel.

Rice, the staple food of Sri Lanka, provides almost 42 per cent of the average energy consumed, while wheat, which is entirely imported and locally milled, contributes 15 per cent to calorie consumption. Coconuts provide the only source of fat and contribute 20 per cent to calorie consumption, besides meeting protein needs. Maize, millets, and other cereals contribute about 2 per cent of the calorie requirements. Protein supplies of animal origin (fish, meat, milk, and eggs) are estimated at 17 per cent of the total protein intake.

Food consumption on an average over the years has increased to the minimum requirement of 2,200 calories per head per day, although considerable ups and downs have been recorded during the 1960s and 1970s chiefly due to variations in rice, wheat, and sugar availability.

TABLE. Per Capita Cereal Availability and Import Dependency Ratio

Period	Average cereal availability (grams per capita per day)	Import dependency ratio (%)
1960-1964	352	52.2
1965-1969	369	51 8
1970-1974	362	45.8
1975-1979	381	48.4
1960-1979	366	48.6

These low-cost staples are extremely important and constitute a valuable source of energy supply during food shortages, adverse weather conditions and droughts, and also in times of calamities.

Even more important is the fact that these kinds of food commody are the only source of food supply that could be reached by the very lowest deciles of the income groups who are unable to increase their purchasing power or market activities, irrespective of economic growth or increases in GNP per capita.

TABLE. Availability of Roots and Tubers, 1960-1979

Year	Roots, tubers, and other starchy foods (tons)	Loss (tons)	Imports (tons)	Edible portion (tons)	Grains per capita per day (g)
1976	961.91	287.68	-	674.23	135
1977	713.40	215.64	0.14	497 90	98
1978	748.16	225.58	1.29	523.87	101
1979	721.09	219.28	7.61	509.42	96

TABLE. Production of Food Crops, 1975-1979 (in tonnes)

	1975	1976	1977	1978	1979
Cereal grams					
Maize	34,552	31,196	41,995	35,206	26,033
Sorghum	6,477	1,662	2,002	544	183
Kurakkan	20,555	15,514	23,335	14,610	10,758
Meneri	490	838	1,077	518	102
Pulses and legumes					
Black gram	1,046	2,362	11,705	8,646	6,043
Cow-pea	7,560	12,028	21,323	22,517	18,772
Green gram	6,324	5,122	7,795	8,400	9,681
Lentil (dhal)	193	41	–	38	53
Soybean	1,162	747	1,112	2,870	1,328
Roots and tubers					
Potatoes	27,302	38,753	33,422	38,453	52,468
Manioc	766,772	683,932	544,933	497,678	364,264
Nuts and oil seeds					
Gingelly	10,748	8,634	7,419	19,691	17,176
Groundnuts	7,604	6,083	5,697	7,475	5,460

Calorie consumption by various income deciles of the population is shown in table below, from which it is evident that the lowest three or four deciles are substantially deficient in calorie consumption. It is among the low-income categories of the population who live in the rural, estate, and urban slums that protein-energy malnutrition has increased during recent times.

NUTRITION SECTOR

It is in this context that the interactions of the nutrition sector to agricultural production on the one hand and food science and low-cost technology development on the other become clearly discernible. A recent study done by the Food and Nutrition Policy

Planning Division for the Economic and Social Commission for Asia and the Pacific (ESCAP) observed that the lowest decile will have to enhance income levels by at least 40 per cent while the second and third deciles have to obtain additional financial support by as much as 20 per cent and 10 per cent in order to meet basic calorie consumption needs.

This position has to be matched against deteriorating nutritional status (FNPPC 1980; CDC 1975/76) in the estate and village sectors where 40 per cent of the pre-school population under 72 months of age are suffering from either chronic undernutrition (wasting) or acute undernutrition (stunting) or both (concurrent chronic and acute undernutrition). Incidence of chronic undernutrition has remained close to the national average of 34.6 per cent, while the rate recorded for acute undernutrition has risen sharply over the national average of 5.6 per cent, particularly in the Mahaweli-H area. The occurrence was comparatively higher in settlers under two years in settlement since this group faces such severe handicaps as lack of finances, low incomes, poor farming experiences, illiteracy, environmental problems, and absence of cohesiveness with neighbouring communities. A higher incidence of nutrition anaemia too has been reported in recent years.

Having been alerted to the crisis, the Government has taken positive measures in relation to food subsidization and the restructuring of wage policies. The general purpose Food Stamp Scheme has benefited over 72 million persons through a cash subsidy in the form of food stamps to procure a basket of nutrition-oriented food commodities (rice, flour/cereal, sugar, milk foods, dry fish, and pulses). Households earning a monthly income below Rs 300 and having an average family size of five to six persons are eligible under this scheme for state assistance. The Government's commitment to this scheme is estimated at Rs 1,550 million, which is approximately 20 per cent of its current expenditure bill. The scheme, though beset with specific constraints, can be considered the best policy option available to Government at the present time when viewed against the problem of adverse trade deficits and the gradual transition from less welfare-oriented policies towards a more dynamic economic development aimed at maximum resource utilization and employment generation.

In addition to this strategy the Government has from time to time provided moderate wage increases in both the private and public sectors in order to compensate for highly inflationary situations, which, if unchecked, could further increase the calorie deficits of the lowest deciles of the population. It could be ideal though less practical to determine a nutrition-based poverty line and define calorie-based wage rates for the village, estate, and low-income urban sectors.

While these two approaches would constitute the Government's major strategy to offset income erosion with a view to enable better food consumption, it is vitally essential to maintain maximum levels of efficiency of the ongoing nutrition interventions both curative (Thriposha and School Biscuit-Feeding Projects) and preventive. The Thriposha programme has reached almost 60 per cent of the target beneficiaries (pre-schoolers, lactating and pregnant mothers) while the School Biscuit-Feeding Project is designed to cover 1.25 million schoolchildren during 1980, a significant percentage of this category being malnourished.

Moreover, there is food aid provided mainly as flour under the United States PL 480 commitments, while food rations like wheat flour, pulses, legumes, and skim milk powder supplies are provided under the World Food Programme to labour working in small-scale rural infrastructure development projects. Labour wages are very marginal in settlement schemes and the workers are unable to meet their basic livelihood needs. Also there is public assistance provided through the Social Services Department for about 350,000 malnourished, destitute, and invalid persons who have absolutely no means and are incapable of earning a living.

FOOD SCIENCE SECTOR

In Sri Lanka the development of the food science sector has not kept up with the challenging demands of a food industry related to agriculture and the nutrition needs of poorer communities. The development of rice, flour, and sugar-processing technologies have taken pride of place over the other agro-industrial crops. The rice-processing industry has been localized to less sophisticated technologies, unlike the wheat and sugar industries where the operations are centralized and highly capital-intensive. This is particularly so of the wheat milling industry.

The rice mills, numbering over 1,400, are scattered throughout the districts although considerable discrepancies are noticeable in respect of mill locations. In the past, parboiling and milling of rice were geared to produce rice varieties (70 to 90 per cent polish) that were more retentive of vitamins and nutrients and hence more nutritious than the highly polished rice varieties that are currently produced. The consumption of highly polished wheat and rice varieties in the developed world has been associated with the higher incidence of certain types of bowel disorders and even bowel cancer due to a lower crude fibre intake.

The losses arising from post-harvest rice losses still continue to be in the region of 20 to 25 per cent of total yields. Grain pest infestations and damage, disease incidence due to poor seed sanitation, and moist conditions of storage are some of the major reasons attributed for these losses.

A Rice Processing and Development Centre has been set up by the Food and Agriculture Organization (FAO) to assist both proprietors of the private and state-owned mills in fulfilling equipment, milling, and training needs. Rice processing technology should be extended to include also the processing and production of rice bran oil as it is an unsaturated fatty oil and better for health.

The entire wheat requirements (450,000 tons/annum) of Sri Lanka is imported. The processing technology is highly sophisticated and product quality is high. Prima Ltd is a joint Sri Lanka-Singapore project that produces almost two-thirds of the country's flour requirements, while the balance is manufactured by the State Flour Milling Corporation. The country imports almost 80 per cent of its sugar requirements (150,000 tons). Sugar production is undertaken by major sugar complexes (Kantalai, Hingurana, Galoya, and Udawalawe). Two other factory complexes are being established in Sevanagala and Moneragala. These industrial units have installations of high capacity (2,500 tons of sugar-cane per day) and are provided with cane supplies by state-owned nuclear plantations and the small-holdings sugar sector. The production of jaggery (39,629 kg) and sugar-cane syrup (129,922 litres) from coconut toddy, kitul, and palymarah, and from cane is also being done by the small industrial sector. But this group of entrepreneurs have always been subject to the vagaries of sugar pricing policies and have been a neglected sector.

Post-harvest losses in other field crops in Sri Lanka have not been examined in depth as have been rice and sugar. Undoubtedly post-harvest losses in this area are likely to be even greater than in rice, since the field crop subsector has not been given the priority it deserves by institutions involved in food technological research and development. The Food Technology Unit of the Department of Agriculture and its Dryzone Research Institute have certainly made some valuable contributions in this direction but much remains to be done specifically in terms of village-level food processing.

This is an area of immediate concern specially considering the tremendous investments made in agricultural development and expansion of the village-level agriculture infrastructure, an area that has received a tremendous impetus from the present Government. Village-level food processing becomes critical in the context of food pricing, conservation, quality control, and availability of food to poorer groups, since it is evident that low-cost staples, coupled with appropriate food and nutrition technologies, would provide a partial answer to the real needs of the low-income, malnourished, and atrisk groups of the population.

Another aspect of food science that has to be emphasized is the area of food surveillance and quality control. Sri Lanka has a very effective food law based on a recent enactment of Parliament but the infrastructure and technical personnel required to enforce and implement the law is seriously lacking. This must be remedied if we are to improve the quality and nutritional level of foods and beverages consumed, more so in view of the production of substandard foods by local manufacturers and the liberalization of food imports.

FUTURE PROSPECTS FOR STRENGTHENING THE INTERPHASES OF AGRICULTURE, NUTRITION, AND FOOD SCIENCES

The immediate goal should be to reach self-sufficiency in rice production as early as possible. Being the staple food, rice would significantly contribute towards per capita food availability in the country. Moreover, the resulting foreign exchange savings accruing from import substitution could be utilized to purchase other nutritious and cheaper foods to reduce existing deficiencies in this area of production.

The whole gamut of the CGRPT group should receive high priority at the national and international levels. For instance, chronically food-deficit countries will have to rely heavily on imports of low-cost food staples in the near future in view of the incompatibility of world market food prices of basic commodities like rice, wheat, and sugar.

Even if Sri Lanka reaches self-sufficiency in rice production a situation is still likely to exist where specific income groups will be unable to procure the basic rice requirements, because rice technology both in terms of production and processing is oriented to capital-intensive inputs such as fertilizer, agro-chemicals, and machinery.

It is thus necessary to preserve rice varieties that are less sophisticated in genetic make-up, less demanding on local resources, and cheaper to produce for the benefit of subsistence farmers, low-income agricultural labourers, suburban labourers, and the class of wage-earners depending on small-scale cottage industries for their livelihood.

Further, agricultural effort should be enhanced through more credit, better rural markets, and extension , and the development of low-cost technologies for the CGRPT group, most of which are quite rich in calories and some in protein. This class of food crop requires a smaller land area and a lesser application of water than do the capital-intensive cropping patterns of other field crops. The regional specialization of nutritive food crops becomes imperative in the context of the strategy argued in this paper. The research and extension services required for this purpose have been established by the setting up of eight Regional Research Stations in the different agro-ecological zones of Sri Lanka, while smaller units are being established elsewhere.

To sustain the type of research and technology development effort required for the promotion of low-cost food staples, it is necessary that the countries of the Asian region should not only develop an effective agro-based food technology network for the transfer of appropriate food technologies, but also explore the possibilities of obtaining funds from the region and other technical assistance towards achieving this goal.

The weaning food industry would serve as a good example to illustrate the use of locally grown cheap foods in the formulation of infant cereal and protein foods that are imported at high cost and are beyond the reach of the malnourished poor. In Sri Lanka work on this aspect is of a very preliminary nature. Laboratory trials have commenced on a few locally formulated weaning foods using different combinations of rice, maize, soya, and green gram to meet specific age-group requirements.

Finally, the interphases linking agriculture, nutrition, and food science can best be understood by providing appropriate levels of training to policy-makers, planners, administrators, trainers, extension personnel, etc., who are involved in work relating to these specialized fields. Further, such training both at the local and regional levels could establish better interaction and co-ordination among these three disciplines. In Sri Lanka nutrition is still conceived in some circles as a purely health-related function, exclusively directed towards primary and maternal health care. Hence a proper and unified communications strategy should be directed not only to the target groups but also to the trainers, to indicate the broad relationships that exist among the various interphases of these three major disciplines.

- Assessment of post-harvest losses in cereals, legumes, oilseeds, fruits, and vegetables in the rural areas.
- Survey of traditional methods of storage of grain and fruit crops, and recommendation of the most effective ones for adaptation.

- Improvement in traditional solar drying methods.
- Designing of low-cost fish preservation and transport techniques.
- Research on the most simple and inexpensive techniques of increasing the shelf-life of fruits and vegetables.
- Rapid parboiling and drying methods for rice.
- Simple inexpensive methods for detoxifying lathyrus (Khesari).
- Designing of economic and effective small-scale milling techniques for pulses.
- Finding use for various agro-wastes.

CONSUMPTION

Nutrition Surveys

Malnutrition is undoubtedly the biggest problem affecting public health in Bangladesh today. Some 80 per cent of the population lives below the poverty line and is therefore not able to afford a balanced diet. Two nutrition surveys, conducted in 1962/64 and 1975/76 (United States 1966; Anon 1978), have confirmed the existence of widespread malnutrition among the rural and urban poor. Salient features are as follows:

TABLE. Per Capita Food Consumption in 1962/64 and 1975/76 (in grams/per capita per/day)

Source	**1962/1964** survey	**1975/1976** survey	**Decrease** (%)
Cereals	528	523	1
Roots and tubers	55.5	52.3	6
Fats and oils	6.6	3.8	42
Pulses and nuts	27.9	23.8	15
Sugar and sweets	7.5	7.3	3
Milk and milk products	17.3	16.8	3
Meat	6.4	3.8	41
Fish	33.3	22.8	31
Vegetables	134.6	125.7	7
Fruits	20.6	10.2	50

TABLE. Per Capita Nutrient Intake during 1962/1964 and 1975/1976

Nutrient	1962/1964 survey	1975/1976 survey	Rural requirement[a]	Percentage of requirement
Calorie (kcal cal)	2,251	2,094	2,248	93
Carbohydrate	476	439	-	-
Protein (g)	57 5	58.5	45.3	129
Fat (g)	15.4	12.2	-	-
Calcium (mg)	304	305	-	68
Iron (mg)	97	22.2	-	292
Vitamin A (IU)	1,590	730	2,013	36
Thiamine (mg)	1.47	1.65	0.90	183
Riboflavin (mg)	0.53	0.87	1.35	64
Niacin (mg)	22.8	22.21	14.84	150
Vitamin C (mg)	39.8	9.5	26.0	37

a. Computed on the basis of 1974 WHO recommendation (Who 1974).

Per capita availability of all the food items was reduced during the intervening period, the least reduction being in cereals and the most in fruits. Calorie availability was reduced by 7 per cent in 12 years. Availability of fats and oils and animal products was reduced markedly. Reduced animal products critically affected the availability of calcium and vitamin A in the diet.

TABLE. Relationship between Cultivable Land Holding per Family and Nutrient Intake of Family Members (per capita per day)[a]

Landholding (acres)	House holds (no.)	Calories (kcal)	Protein (g)	Fat (g)	Niacin (mg)	Calcium (mg)
0	175	1,925	53	11	20	239
0.01-0.49	127	1,924	53	10	21	256
0.50-0.99	110	2,035	58	11	22	329
1.00-2.99	160	2,193	62	13	23	331
3.00 and over	90	2,375	68	14	25	386
WHO standard	-	2,122	39	16	12	503

Source: University of Dacca 1978b. Figures for rural Bangladesh 1975/76.

The 1976 survey showed the following features:

- Protein deficiency was evident in 60 per cent of families, and 30 per cent were calorie-deficient.
- A total of 81 per cent of households were deficient in calcium intake (this was the same in the two surveys).
- Households were deficient in vitamin A (89 per cent), riboflavin (85 per cent), and vitamin C (93 per cent).
- Cereal intake in adult males was significantly higher for children and females of all categories.
- All children up to 15 years of age, and adults between 20 and 39 years of age did not get enough calories.
- All children up to six years of age, and lactating as well as pregnant mothers, were deficient in proteins and calories.
- Vitamin A and C intake did not exceed 70 and 54 per cent of requirement, respectively, for any agegroup.
- Sixteen per cent of children were found to be wasted and stunted (concurrent acute and chronic malnutrition), 54 per cent had chronic undernutrition, and 6 per cent were wasted. Only 20 per cent had normal height and weight. Girls were found to be more malnourished than boys.
- About 0.8 million of the population was found to suffer from night blindness.
- Ten per cent of children had angular stomatitis.
- Despite more iron than required being present in the diet, 70 per cent of the population were anaemic; 82 per cent of children had haemoglobin levels below normal.
- Income showed a positive relationship with the intake of cereals, pulses, and fish, but not with vegetables.

Average intake of all families having less than three acres of land is lacking in calories, calcium, vitamin A, riboflavin, and vitamin C..

Total food intake is more in the first six months of the year than in the last six months. Inadequacy, maldistribution, and lack of knowledge about food are responsible for malnutrition.

Lack of hygiene and sanitation further exaggerates the situation. All these factors are linked with poverty.

National Nutrition Council

In 1977, a sub-committee appointed by the National Nutrition Council made the following important recommendations:

- Nutrition must receive topmost priority in the national development plan.
- Being a multidisciplinary area, nutrition planning calls for an integrated approach.
- Nutrition education, both formal and informal, has an important role to play in improving the nutritional status of people.
- Production of all food commodities needs to be increased by harvesting more calories per unit area in the least time and with the minimum inputs.
- Post-harvest losses should be reduced to the minimum.
- Fisheries, poultry, and livestock management needs to be improved.
- Environment and health care need to be improved to enhance absorption and retention.
- Employment and integrated rural development must receive importance.
- Special feeding programmes, based on local ingredients, should be launched for the under-five age-group.
- A permanent secretariat for the Nutrition Council and a Parliamentary Committee on
- Nutrition should be established.

Nutrition in the SFYP

The basic needs method has been used to estimate the food demands projected in the SFYP. Unfortunately, population figures used to compute the requirements are underestimates, as was proved by the recently completed census. On the production side nutrition needs have been simply translated into agricultural production figures without considering feasibility. While the 1975/76 nutrition survey indicated that the availability of all but carbohydrate (starch) crops had reduced markedly since 1964, the SFYP continues to lay emphasis on increased production of cereals with relatively less resource allocation on other crops. This bias, coupled with unequal distribution of food, is likely to lead to further malnutrition in the vulnerable classes of the population. Furthermore, seasonal and regional differences in the availability of food ingredients need to be taken into consideration in preparing the food balance sheets and in formulating agricultural strategies for affected areas and populations (Kaul 1977b).

High incidence of roundworm (Ascaris lumbricoides), hookworm (Anclystoma duodenable), and whipworm (Trichuris trichiura) has been reported in Bangladesh. The population affected may range from 50 to 80 per cent, depending on the location and living conditions. Roundworm is the most common parasite found. The incidence of anaemia, despite a satisfactory average iron intake of 22 mg per day, is a direct consequence of parasitism. Similarly, other parasites and communicable diseases lead to serious malnutrition, particularly in children. This aspect of health care must be taken into consideration in nutrition planning (Ahmed 1981).

Several million children and women located in three northern districts of Bangladesh suffer from goitre. These areas, being specifically iodine-deficient, can be effectively attended to through programmes involving the supply of iodized salt and injection of iodized oil (Ahmed 1981).

INTERFACES

A successful food policy, besides production aspects, must take into account distribution, conservation, utilization, and retention of nutrients. Increasing production and productivity is only one aspect of the problem. In Bangladesh a major cause of malnutrition is inability of the poor to purchase adequate supplies of nutritious foods or to retain home-grown food for family consumption. Therefore it is primarily a question of raising incomes. Other factors leading to malnutrition are infection, ignorance, biased intra-family distribution of food, poor health, and sanitation. These factors lead to losses and poor utilization of what little is available. Other policy issues concern pricing, investment, resource allocation on pertinent research areas, and degree of interaction and feedback between agriculture, food, nutrition, and social sciences. Plan projections should not be based exclusively on the "basic needs" principle but on practical feasibility. Tools increasing productivity must simultaneously generate employment, purchasing power, and overall well-being of the society.

Increasing the production of cereals is the easiest of all challenges. There is considerable scope for increasing rice and wheat production. Maize has excellent prospects in Bangladesh. By shifting from rice to wheat and maize, protein output per unit area is likely to increase. The prospect of harvesting ample calories from tuber crops, particularly potato and sweet potato, has been realized only recently. Tuber crops had not received much attention in the past.

Increasing the production of pulses is an urgent need. Introduction of new varieties of traditional pulses from neighbouring countries has brightened the prospects for production of these crops. The prospects for soybean and winged bean use are bleak as long as their acceptability is not increased in the rural areas. Processing and preparation of soybean for the urban market remains an important objective of the Food Sciences Division of the BCSIR, but the production and agronomy of this crop remain divorced from industry. The result has been very slow progress in soybean cultivation, despite sizeable investment on research. The seed viability and post-harvest preservation aspects were completely missed in earlier planning. Conventional pulses have by necessity to be encouraged. Postharvest preservation must receive utmost attention. The complementary role that pulses play in the cereal-based diet needs recognition and demonstration through feeding experiments.

The acute shortage of fats in the diet has been realized. There are bright prospects of increasing the production and productivity of oilseed crops in Bangladesh. Besides brassica, groundnut, sunflower, sesame, and safflower have immediate prospects. Extraction processes are rather primitive but, in the interest of rural employment, these might be left undisturbed.

The acute shortage of animal proteins is hardly necessary in Bangladesh. Fisheries, both inland and marine, goat husbandry, backyard poultry, and animal husbandry can all be organized in a manner that ensures higher production and employment generation at the same time. Processing technology and improvement of storage methods could be improved considerably even with the available know-how. Fish production can be increased twofold to threefold in five years' time.

Augmentation of vitamins and minerals in the diet through increased production and consumption of fruits and vegetables is possible. However, the landless, who represent more than 50 per cent of the rural population, can have access to these food items only if their earning capacity is increased. Yield potential can definitely be increased by a factor of two but preservation and processing require sizeable capital inputs. On the nutritional front, mass education on the value of fruits and vegetables needs to be emphasized.

This is one food item that was found not to be correlated with income. Several local fruit and vegetable species are excellent sources of vitamins, but for lack of knowledge of their value, the same have not been utilized effectively. Food sciences have not exploited this wealth so far.

A good deal of food is wasted in Bangladesh through pests in the field, during storage, and in the human gut. While integrated pest- and disease-control mechanisms have to be developed for the field, effective storage and processing methodology needs to be perfected to prevent post-harvest losses. Education and improvement of sanitation and health care could avoid considerable losses during consumption.

Improving the nutritional status of the Banglalee population does not merely involve increasing production through harvesting of more food per unit area in less time with minimum expenditure. It calls for perfecting a production machanism that gives opportunity to the producers to consume the food they produce and educating them on methods that would avoid losses at all levels of production, processing, and consumption. Such an approach demands integration of hitherto separate policies of food production distribution, and consumption. The concept of the integrated Crop-Livestock-Energy farm deserves to be explored.

Agriculture provides 55 per cent of the GNP, 85 per cent of the employment, and 80 per cent of the export earnings of Bangladesh. Yet more than 70 per cent of the population suffer from second- and third-degree malnutrition. Landless families constitute the bulk of the nutritionally disadvantaged. Only a policy of drastic land redistribution might promote both production and equity. Considerable scope exists for increasing the production of all food commodities. Prevention of post-harvest losses, which may not exceed 5-7 per cent, should not be treated as a third soft option of food policy. Food scientists have yet to address themselves to the rural problems of Bangladesh. Infection, parasites, and poor sanitation add to rural malnutrition. Lack of education on the food value of local fruits and vegetables is unfortunate. In Bangladesh a production mechanism needs to be perfected that gives opportunity to the producers to consume the food they produce and that will educate them on methods to prevent losses at all levels of production, processing, and consumption. Such an approach calls for integration of hitherto separate policies of food production, distribution, and consumption. The concept of an integrated "Crop-Livestock-Energy" model family farm is discussed.

14

Interaction Between Agriculture, Nutrition, and Food Science in Nepal

Agricultural production is not keeping pace with the population increase. Post-harvest losses of food grains and perishable foods like fruits and vegetables are thought to be around 10-15 and 20 35 per cent respectively, and whatever is saved loses some of its qualitative value due to faulty storage and processing practices.

Food availability, specially among weaker sections of the population, is below normal requirement and undernutrition in terms of protein/calorie, iron, carotene, and iodine deficiency is rampant in Nepal. Increased food production is possible through increased exploitation of abundant water resources and greater application of agricultural inputs and technologies. Increase in agricultural production will increase food availability and improve economic status, the two most important prerequisites of nutritional improvement.

Application of food science and technology will not only help in increasing food availability per unit area through high food value giving seed selection, but will also help to conserve, reduce food waste, and improve the quality of produced food. Nutritionists have to draw the attention of agriculturists and food scientists to the nutritional needs of the population and work with them for the quantitative and qualitative fulfilment of nutritional needs. To sum up, agriculture, food science, and nutrition jointly offer great hopes for nutritional improvement.

INTRODUCTION

Nepal is predominantly a mountainous country with a total land area of 140,000 km^2 of which about 14.1 per cent are currently under cultivation. According to the Nepal census of 1971, the population was 11.3 million. Agro-climatically the country is divided from north to south into a sparsely populated mountain area with an alpine climate, a hill region accommodating about two-thirds of the population producing one-third of the total tropical and subtropical foods, and a flat terai region extending from west Bengal to western Uttar Pradesh in India, where one-third of the people live and produce two-thirds of the tropical foods. Due to variation in rainfall from east to west, the country is further subdivided into four development regions, each containing mountains, hills, and terai.

About 90 per cent of the population are engaged in farming and related fields contributing 62 per cent of national production and 80 per cent of the export value. Thus basic information shows that agriculture is responsible not only for the food supply, but it contributes heavily to economic development.

AGRICULTURE

Food Production

The population has been increasing at 2.12 per cent, but food production increased by only 0.79 per cent annually during the period 1964/65 to 1978/79, showing that food availability was decreasing. It is extrapolated that cereal grain production will increase at the rate of 1 per cent provided present increased production efforts are maintained, but the population will increase at a 2.4 to 2.5 per cent rate. The National Planning Commission (1980) has outlined the reasons for low agricultural productivity.

1. Low irrigation facilities due to under-exploitation of abundant water resources.
2. High rate of soil erosion due to a high rate of deforestation;
3. Reduction in anticipated application of agricultural inputs like fertilizer, improved seeds, etc.

4. Lack of co-ordination between agricultural extension, agricultural credit, agricultural input, and cooperatives.
5. Under-exploitation of available technologies to boost production.

Agriculture was for the first time officially related to food and nutrition by the National Planning Commission when it modified its long-term policies in the following directions.

To increase production of food grains, fruits and vegetables, and livestock products to fulfil the required food and nutritional needs of the growing population.

To increase agricultural production so as to generate the capital requirement for economic development and to maintain an export requirement.

To provide raw material for agro-based industries and to increase employment opportunities.

To increase agricultural production at a higher rate in the hill region to remove disparity between regions.

Food Availability

Total food-grain production is partly accounted for by post-harvest losses, seed, and feed requirements, etc. The remaining available food when divided by population gives the total availability of food grain for consumption; this was 179 kg per head per year in 1964/65, but decreased to 138 kg during 1979/80. The total estimated consumable available foods in the hill region during 1979/80, computed against the population of the same year, shows that the per capita calorie availability per day was only in the average range of 1,759 kcal.

Considering the economic variation and distribution disparity between communities and within families, certain parts of the population must be trying to survive below their normal metabolic requirement.

This short supply of food must be the leading causative factor of undernutrition. Disturbed by this problem, the National Planning Commission (19801 has very clearly specified as the first and foremost objective of the sixth agricultural plan that present-day estimated availability of 2,181 calories and 56.6

grams of protein must be increased to 2,266 calories and 59.8 grams of protein by 1985. This drastic change in policy clearly shows the realization of the importance of nutrition in overall national development, and heavy reliance on agriculture and food science to solve this problem.

FOOD SCIENCE AND TECHNOLOGY

It is thought that at least 10 to 15 per cent of food grains and 20 to 35 per cent of perishable foods like fruits and vegetables do not reach the consumer's plate due to post-harvest losses. When these food-grain losses are computed against total production, 241,000 to 362,000 tonnes of food grains must have been lost in one year. At the same time there is an appreciable proportion of qualitative loss as well. Preliminary food contamination observations show that this is not only reducing the biological value of food, but also turning food into unseen poisons.

In the interaction between food science and technology during the improved rice varieties development programme, it was noted that the husk percentage in different varieties of rice varies from 21 to 27.25 and on one-minute polishing the bran percentage varied from 4.3 to 10.22 (Bhandari 1979/80). Similarly the percentage of cracked grain is also a characteristic of each variety. When these varieties were milled in the Food Testing Laboratory mill, the milling yield varied from 66.3 to 74.18 per cent.

This shows that the interaction between the agricultural production potential and food science and technology will not only make more food available, but that the higher economic return to the farmer in turn will encourage the farmer towards increased production. In another unpublished study of the Food Research Laboratory (1976) it was noted with concern that the milling recovery from a huller mill is only around 59 to 62 per cent whereas the average recovery rate from modern rice mills is 66.5 per cent.

Similarly, application of food science and technology will help to increase food supply through improved agricultural and industrial by-products utilization.

TABLE. Husk, Bran, Rice, and Head Rice Percentage Recovery From Different Varieties of Paddy

Varietal name	Milling percentages			
	Husk	Bran	Rice	Head rice
IET 4183	24.25	9.05	66.70	(49.39)
IR 2071-625-1	24.00	6.46	69.54	36.06
BG 34-8	22.75	7.22	70.03	81.72
NR 6-5-46-45	23.00	6.82	70.18	64.90
P 23-C-19	25.50	8.20	(66.30)	51.86
IR 2061-628 (Laxmi)	*27.75*[a]	*10.22*	62.03	51.86
CHANDINA	23.23	5.30	71.45	72.06
IR 3707-1172	24.54	4.61	70.89	73.11
IR 36	21.50	5.07	72.53	71.42
PUSA 33-30-3-3	27.25	6.26	66.49	75.82
BG 94-1	23.25	4.61	72.14	83.24
IET 2936-DURGA	24.00	4.79	71.21	82.79
IR 2797-125-3	24.25	6.97	68.78	52.20
B 1665-B-MR-7-5-15	21.25	5.04	73.71	*94.02*
B 805-DMR-16	(21.00)	4.82	*74.18*	93.38
IR 5677-35-4-2-3	23.00	4.85	72.15	87.63
B 44-B-50-2-2-5	21.25	4.73	74.02	92.94
IR 2071-58-6-5-6	23.25	(4.30)	72.45	81.09
ER 51-91-7	22.20	4.65	72.85	92.37
IR 3478-97-2-3	21.50	5.18	73.32	84.42

a. Figures in italics represent highest values

It is known that a suitable combination of foods improve nutritional value. This knowledge when integrated with food science and technology, food availability, food preference, and food habits could produce products to improve the nutrition of the population.

NUTRITIONAL STATUS

A limited number of studies on the extent of undernutrition have been made. The prevalence of oedema in children attending clinics in Kathmandu Valley was about 5 per cent (Van Dijk 1968). In another study in the same area about 20 per cent of children 10-24 months of age were found to be below 80 per cent of the Stuart Meredith reference median (Ziegler and Ziegler 1976). Another study by Pourbaix (1974) in 17 clinics and villages in different parts of Nepal showed that 71 and 19 per cent of the pre-school children surveyed were below 80 per cent of the Stuart-Stevenson reference medians for weight-for-age and weight-for-height respectively.

Another anthropometric survey (Brink et al. 1976) employed population-proportionate sampling throughout rural Nepal, excluding 14 town areas. In addition 486 children of well-to-do families of Kathmandu and Patan were selected for a "special group" survey. These children were included to gather information on families who can afford to feed their children well if they want to. There was a high proportion of normal children in the terai area ($P < .001$) and considerably more stunted children in the hiliy area ($P < .001$).

TABLE. Percentage Distribution of Children 6-71 Months of Age by Waterlow Categories(a)

Geographical divisions	Normal	Stunted only	Wasted only	Wasted and stunted
Far west	41.6	51.2	2.7	4.5
West	42.6	51.7	2.0	3.7
Central	46.7	45.6	3.6	4.1
East	48.4	45.8	3.0	2.8
Terrain				
Hill	42.1	52.1	2.2	3.6
Terai	50.5	40.8	4.4	4.3
Total for rural Nepal	45.0	48.3	2.9	3.8

a. Percentages based on weighted data.

Prevalence of malnutrition in different age-groups showed that Nepalese children have reasonably good physical growth after birth, but a marked increase in stunting took place with increase in age. The highest percentage of wasting was found in the second year of life, reflecting the problem of transition from breast milk to external food. A much higher percentage of protein-energy undernutrition is a general problem of rural

Nepal. Nearly 4 per cent of the children are wasted and stunted, and 50 per cent have a significant deficit in linear growth.

TABLE. Age-specific Percentage Distribution in Waterlow Categories. Nepal Survey Population (NSP) and "Special Group" (SG)

Age-group	Normal	Stunted only	Wasted only	Wasted and stunted
(months)	NSP (SG)	NSP (SG)	NSP (SG)	NSP (SG)
6-11	70.9 (83)	21.0 (17)	5.9 Nil	2.2 Nil
12-23	45.6 (86)	39.1 (11)	6.4 (1)	8.9 (1)
24-35	44.1(82)	46.8 (13)	3.8 (3)	5.3 (2)
36-47	40.4 (85)	56.2 (14)	0.8 Nil	2.6 (1)
48-59	36.3 (73)	61.4 (24)	0.7 (2)	1.6 Nil
60-71	44.1 (75)	54.2 (25)	0.7 Nil	1.0 Nil
TOTAL	45.0 (85)	48.3 (18)	2.9 (1)	3.8 (1)

In the case of the special group less than one-fifth of the children have a significant deficit in linear growth and less than 1 per cent are severely stunted and wasted. It is significant that a growth pattern currently displayed by more affluent Nepalese is potentially attainable, and this should be the target for long-term nutritional programmes.

TABLE. Mean Haemoglobin Values[a] for Rural Nepal and Percentage Below Normal for Age

Age-group (months)	Haemoglobin in rural Nepal children	Standard deviation	Percentage with haemoglobin values	
	g/l		< 100 g/l	100-109 g/l
6-11	110	15	19	30
12-23	112	17	20	21
24-35	114	16	17	22
36-47	117	17	9	16
48-59	120	19	9	14
60-71	121	17	6	13
TOTAL	117	17	12	18

a. Corrected for altitude.

If we accept criteria for the diagnosis of anaemia on the basis of a haemoglobin level of less than 100 grams per litre for children 6-23 months and less than 110 grams per litre for children 24-71 months, then anaemia is most prevalent in children in the 24-35 month age-group. In general anaemia is an important problem among rural children of Nepal and probably reflects iron deficiency due mainly to an inadequate dietary intake.

NEW PERSPECTIVES

Of the three basic needs of fellow human beings, shelter and clothing are still considered secondary. Food has always been the prime need and the centre of great social and political changes.

Malthus warned that while food supplies increase at an arithmetic rate, the population increases at a geometric rate and if this is true the population will outstrip available food supply. But even from neolithic days man was warned about this situation. Later, the increase in cultivable land, improved method of farming, application of fertilizer and irrigation, and lately the Green Revolution, has allowed production to keep pace with demand. Newer developments and understanding in the fields of nutrition and food science and technology reveal that purely vegetarian diets viewed against the increasing prevalence of obesity and hyperlipidoemia, can promise good health. Moreover, a vegetarian diet is about five times more economic to produce than one which includes animal foods. Further, newer knowledge in food science and technology promises to increase and improve the quality and availability of food through seed variety development, better processing and preservation techniques and by-product utilization.

Nutrition is still a young science and its sister subjects of food science and technology are even newer. The subjects of nutrition and food science and technology are closely linked and can even be regarded as different faces of the same coin. Existence of food science and nutrition is unthinkable in the absence of food that is the primary produce of agriculture. Certainly one cannot function in the absence of either of the others.

Ignorance and taboos restrict the consumption even of certain foods available at hand, while at the same time traditional foods related to religion and custom are contributing favourably towards improvement of nutrition. Evaluation of these traditional foods, and the application of food science followed by popularization for more frequent consumption can help improve nutrition.

CONCLUSION

Agricultural production is not keeping pace with population increase. Post-harvest losses of food grains and perishable foods like fruits and vegetables are thought to be around 10-15 and 20-35 per cent respectively, and whatever is saved loses some of its qualitative value due to faulty storage and processing practices. Food availability, specially for the more vulnerable sections of the population, is below the normal requirement, and undernutrition in terms of proteins and calories, iron, carotene, and iodine is rampant in Nepal. Increased food production is possible through increased exploitation of abundant water resources and greater application of agricultural inputs and technologies. Increase in agricultural production will increase food availability and improve economic status, the two most important prerequisites of nutritonal improvement. Application of food science and technology will not only help in increasing food availability and food value per unit area, but will also help to conserve food, reduce food waste, and improve the quality of food production. Nutritionists should draw the attention of agriculturists and food scientists to the nutritional needs of the population and work with them for the quantitative and qualitative fulfilment of nutritional needs. To sum up, agriculture, food science, and nutrition jointly offer great hopes for the nutritional improvement of the people.

15

Interaction between Nutrition and Agriculture in India

Nutrition is an important environmental factor that influences health, and adequate food intake is a prerequisite for good nutrition. Agricultural production is of crucial importance in ensuring that food needs are met. Countries with a high rate of population growth need to develop suitable agricultural strategies to increase food production. In addition to this quantitative aspect, there has to be concern for the qualitative dimension of food production. The mixed bag of produce should be able to satisfy "recommended allowances" of all nutrients needed for optimal nutrition.

Food production in India has increased substantially over the years. All food grains have, however, not increased uniformly. Production of cereals and millets has increased substantially but not that of pulses and oilseeds. In a country where the inclusion of pulses is important in increasing dietary protein content and for improving protein quality, this is a disturbing trend. At current levels of production, the recommendation of the Indian Council of Medical Research (ICMP) (1968) that a balanced diet should contain 70 grams of pulse, cannot be met. The revised balanced diet (1980) contains only 50 grams and this level can barely be met. Unless special efforts are made to increase pulse production, there is a real risk that recommended dietary allowances of pulses cannot be met.

Another qualitative aspect of food production has been India's efforts to identify, evolve, and propagate food-grain varieties with more-than-average nutrient content. Protein and lysine have received special attention. High protein/high lysine lines of cereals and millets have been identified, but have often been found not to breed true, for reasons not fully understood. Improving protein content and quality of staples was conceived as a method of improving the quality of diets, at a time when habitual Indian diets were considered to be protein-deficient. This concept has changed and the primary bottleneck is now believed to be energy.

Cereal-pulse based diets have been found to be capable of meeting protein needs, when consumed in amounts that satisfy energy needs. The relevance of efforts to improve protein quality, therefore, needs re-evaluation.

At the national level, food production appears to be adequate to meet demands, provided there is equitable distribution. In actual practice many households do not get enough food because of poor purchasing power Among families whose daily per capita income is below Rs 3/-, over one half consumes an energy-deficient diet. A proportion of such households do not get enough protein either - a finding that explains the widespread childhood energy-protein ma/nutrition. The impressive buffer stocks of food grains held in recent years is a reflection of this low consumption. They would disappear should the purchasing capacity improve. Current levels of production under such circumstances would not be enough to build reserves.

Due to increased agricultural production, food-grain import has, normally, all but stopped. What has been achieved in the Indian agricultural situation has been the prevention of serious famines, which occurred in earlier years. But it does not appear to have made much impact on the widespread chronic malnutrition. To be able to reduce chronic malnutrition, increase in food production has to be of a magnitude larger than that seen at present. This alone will not suffice. Food grains have to be within the price range of the great majority. Also, national nutrition policy and national agricultural policy will have to be more compatible.

Nutrition is an important environmental factor that influences health and well-being. Consumption of diets adequate both in quantity and quality is a prerequisite for the maintenance of good nutritional status. Agricultural production that determines food availability is, therefore, an important determinant of food consumption, though not a critical one if food imports can be assured. Self-sufficiency in food production is of particular importance for developing countries, not only because they tend to have high rates of population growth, but also because such countries have malnutrition as a public health problem.

The quantitative aspects of food production are undoubtedly of primary concern, but it cannot be forgotten that the qualitative aspects are extremely important, if optimal nutrition is to be provided. The interphase between agriculture and nutrition, therefore, acquires considerable practical importance. Some aspects of this interphase as they relate to India are briefly presented here.

PRODUCTION OF FOOD GRAINS

Food-grain production in India has risen considerably over the last three decades. The increase has indeed been spectacular - from around 50 million tons in 1951 to over 130 million tons in 1980 representing as it does a 2.5-fold rise. The true significance of this becomes apparent when it is viewed in terms of per capita availability at the national level.

This has jumped from about 350 grams a day to around 470 grams, after allowances are made for food losses and reservation of seeds. The increase in production in the earlier part of the period was of a magnitude greater than the increase in population, and has subsequently kept pace with the population rise. The increased production has been achieved largely through an increase in the yield of grain per hectare, rather than from an increased cultivated area.

The increase in production has been the result of widespread cultivation of new high-yielding varieties coupled with a package of agricultural inputs that permit the genetic expression of their high yield potential. As a result, from cereals, millets, and pulses alone (excluding oilseeds, sugar, roots, and tubers) dietary energy availability has gone up from 1,180 calories per capita daily, to 1,650 calories.

TABLE. Per Capita Food-grain Availability[a] in India: 1951-1980

Year	Total production (million tons)	Net availability (million tons)	Per capita availability (g/day)
1951	50.8	44.4	337
1956	66.9	58.5	392
1961	82.0	72.3	430
1966	72.3	63.3	351
1971	108.4	94.9	473
1975	100.0	87.5	420
1978	126.4	110.0	475
1979	131.4	115.0	482
1980	132.0	115.5	470

a. Excluding oilseeds, sugar, roots, tubers, milk, and milk products
Sources: Bulletin of Food Statistics 1975, 1979; Agricultural Situation in India, 1980

The increase in the production of different types of food grains has, however, not been uniform. While that of wheat has registered a sixfold increase from 6.4 million tons in 1951 to over 35 million tons in 1980, those of rice and millets have been less marked.

This is particularly so since 1961, the increase in rice production between then and 1979 being around 60 per cent and that of millets only 30 per cent. Although the nutrient composition of rice, wheat, and millets does differ in several respects, and though the digestibilities of these staples are different, it is unlikely that this would have demonstrable nutritional significance. What, however, is of considerable nutritional importance is the failure of pulse production to increase between 1961 and the present.

Following a 50 per cent increase between 1951 and 1961, there has been virtually no change over the last two decades. As a consequence, the per capita availability of pulses which stood at 70 grams per day in 1961, has now dropped sharply to about 45 grams. With an increase in cereal-millet availability, the pulse-cereal ratio has shown a marked distortion, falling from about 20 per cent to half that figure at present.

TABLE. Production of Rice, Wheat, and Millets in India: 1951-1979 (million tons)

	1951	1961	1971	1976	1979
Rice	20.6	34.6	42.2	48.7	53.8
Wheat	6.4	11.0	23.8	28.9	35.0
Millets[a]	12.9	20.7	27.8	26.2	27.1

a. Bajra, barley, jowar, maize, ragi, and small millets.

Source: Agricultural Situation in India, 1980.

TABLE. Per Capita Availability of pulses in India: 1951-1980

Year	Total production (million tons)	Net availability (million tons)	Per capita availability (g/day)
1951	8.4	7.3	55
1956	11.0	9.7	69
1961	12.7	11.1	70
1971	11.8	10.3	51
1976	13.1	11.5	51
1980	12.2	10.8	43

a. pulses are important sources of protein, lysine, riboflavin, and trace metals in the Indian diet

Source: Agricultural Situation in India 1980.

TABLE. Per Capita Availability of Cereals and pulses Ratio 1951-1980

	Cereals and millets: g/day	Pulses g/dav	Total g/day	Pulses as ratio of total
1951	282	55	337	19.5
1961	360	70	430	17.5
1971	422	51	473	19.5
1976	403	51	454	12.5
1980	427	43	470	10.7

The habitual Indian diet is predominantly vegetable-based, and foods of animal origin do not usually find a place because of their high cost. The inclusion of pulses in cereal-millet-based diets is critical not only in increasing the protein content, but also in improving the nutritional quality of the protein. pulses are also rich sources of several trace metals, whose essentiality to human beings is no longer in doubt. The present situation has, therefore, to be viewed with considerable concern. The balanced diet recommended by the Indian Council of Medical Research in 1968 contained 70 grams of pulses for an adult subject (Gopalan and Narasinga Rao 1968). A 25 per cent higher production level is needed to support availability at the individual level since allowances have to be made for losses. wastage, and seed purposes. At no time during the last three decades could the production level permit the inclusion of 70 grams of pulses daily.

This figure was perhaps unnecessarily high, and, using linear programming, the least-cost balanced diet recommended recently (ICMR 1981) provides for only 40 grams of pulses per day. To be able to achieve this, the production level has to be over 50 grams a day - a figure that just about matches current production. This is indeed a disturbing situation, and proper corrective measures have to be introduced urgently if the situation is not to become worse. The danger is real, in the face of the very low per capita availability of foods of animal origin, including milk and its products.

TABLE. Per Capita Production Level[a] of Calories and Pulses: g/day

	1951	1956	1961	1966	1971	1976	1980
Cereals	395	470	504	411	570	534	560
Pulses	63	76	78	45	60	57	49

a. Production level needed to meet recommended dietary allowances (RDA): cereals: 490 (1981 RDA), pulses: 53 (1981 RDA); 87 (1968 RDA).

UPGRADING NUTRIENT QUALITY

A qualitative dimension to food production was built into Indian agriculture when attempts were begun years ago, to improve the nutritive value of cereals and millets through genetic engineering, made possible by the discovery of genes that control the concentrations of nutrients. Malnutrition is a major public health problem in India, particularly among pre-school children, and the major reason for this is the consumption of diets that do not provide adequate amounts of several nutrients. Among deficiency diseases, protein-energy malnutrition (PEM) is the most widespread, and based upon the belief then that the diets of children in whom PEM was prevalent, were deficient in protein - both in terms of quantity and quality - emphasis was laid on developing varieties of food grains that had higher-than-average protein content and a higher amount of lysine-the limiting amino acid in cereals.

TABLE. Nutrient Content of Five Sorghum Varieties Grown in Two Locations

Variety	Protein g/100 g		Lysine		Leucine	
			g/16 g			
	D[a]	I[a]	D	I	D	I
CSV-1	7.7	9.8	2.6	1.9	11.7	9.7
CSV-2	7.0	10.7	2.3	1.6	11.5	12.4
CSV-6	7.7	13.1	2.1	-	11.0	13.9
SPV-8	7.6	12.6	2.2	1.5	11.1	12.3
SPV-40	6.0	11.2	2.5	1.9	13.1	13.0

a. D= Dharwar; I = Indore.

Source: National Institute of Nutrition 1977.

It was the discovery of high levels of lysine and tryptophan in the opaque 2 mutant maize, which initiated a search for similar characteristics in other foods, and subsequently varieties of cereals and millets with high protein and high lysine have been identified. When grown under apparently similar agronomic conditions in different parts of the country, they appear not to breed true. For example, when the same five varieties of sorghum were grown in two different locations - Dharwar and Indore - under the Co-ordinated Sorghum Project, those grown in Dharwar had lower amounts of protein than those grown in Indore. The extent of difference was not uniform. values were lower by 21 to 86 per cent, depending upon the variety (table 6). Values for lysine were lower in all varieties grown in Indore, a finding to be expected considering the known inverse relationship between protein and lysine concentrations . But the differences in lysine between the two locations could not all be accounted for by differences in protein alone, The concentration of leucine in the protein also showed changes from marginal to considerable.

TABLE. Nutrient Composition of Red Gram Variety C-11, Grown in Five Locations

Location	Protein g/100 g	Tryptophan 8/16 g N	Methionine
Nagpur	25.2	0.8	0.9
Akola	25.1	0.4	-
Coimbatore	24.0	0.8	1.0
Aurangabad	22.8	0.5	-
Jabalpur	18.1	0.5	1.4

Source: National Institute of Nutrition 1978.

This was also true of the nutrient content of red gram. Three varieties of red gram grown in four different locations showed that the protein content could vary by over 20 per cent without corresponding changes in either leucine or methionine, but with substantial differences in their riboflavin content. Another variety of red gram, C 11, grown in five locations, showed again that protein content could vary by over 28 per cent, the concentration of methionine by 55 per cent, and that of tryptophan by 100 per cent. Differences in the amino acid content seemed not to be related to differences in protein content.

TABLE. Toxin Content (ß-Oxalyl Aspartic Acid in Varieties of Lathyrus sativus Grown in Different Locations

Variety	Range of toxin in nine locations	
	mg/100 g	Fold variation
LSD-1	104-393	3.8
LSD-2	155-600	3.8
LSD-4	166-476	2.9
LSD-5	104-642	6.2
LSD-6	93-590	6.4
P-24	207-559	2.2

Source: National Institute of Nutrition 1978.

Such locational differences do not seem to be limited to nutrient composition. They seem to apply also to the concentration of the unusual amino acid ß-oxalyl aspartic acid, present in Lathyrus sativus and believed to be the toxin responsible for neurolathyrism. They also seem to apply to the potential for supporting the in vitro production of the fungal toxin, aflatoxin.

TABLE. Locational Differences in Aflatoxin Production Potential of Maize Varieties

Variety	Location	Aflatoxin production potential (ppm)
Shakti	Andhra Pradesh	510
	IARI, Delhi	403
DHM-10	Andhra Pradesh	455
	IARI, Delhi	208
B-19	Andhra Pradesh	406
	IARI, Delhi	162
B-23	Andhra Pradesh	480
	IARI, Delhi	63
Ganga-5	Andhra Pradesh	495
	Rajasthan	112

These findings suggest that it would not be easy to predict with any degree of confidence the nutrient composition of a variety of food grain that has a high protein potential. Apart from this practical consideration, improving protein content and protein quality was conceived as a method of improving the quality of the diet at a time when it was believed that protein was the limiting nutrient in Indian diets. This concept has now changed and the primary limiting nutrient is believed to be energy. Cereal-pulse-based diets of the type commonly eaten have a protein-energy ratio which is capable of meeting protein requirements, when consumed in amounts that satisfy energy requirements. It must be stressed that this is not to imply that there is no inadequacy of dietary protein. A fair proportion of subjects do have low protein intakes, but this is almost exclusively because total food intake is so low that both protein and energy become limiting. The relevance of efforts to improve protein quality by genetic manipulation, as a priority measure, therefore, needs re-examination, at least in the Indian context. Whether upgrading the lysine content is relevant may also be questioned, since in cereal-pulse-based diets, the first limiting amino acid is not lysine, but the sulphur-containing amino acids.

TABLE. Adequacy of Protein and Calorie Intake among Families in 1977

Category	Percentage of families in:			
	Karnataka	Maharashtra	West Bengal	Uttar Pradesh
Protein adequate - calorie adequate	69.2	57.8	59.0	60.9
Protein inadequate - calorie inadequate	17.4	14.8	25.9	6.8
Protein adequate - calorie inadequate	11.6	27.4	14.1	32.3
Protein inadequate - calorie adequate	1.8	0.1	1.0	0.1

Source: National Nutrition Monitoring Bureau 1978

Turning to the availability of agricultural produces other than cereals and pulses, particularly edible oils and sugar, the situation is not an encouraging one. The per capita availability of both has gone up: that of edible oils and fats from 7.7 grams per day in 1951 to 13 in 1979, and that of sugar from 8.2 grams per day to 19.8. Though higher now than three decades ago, these are still considerably lower than what the least-cost balanced diet recommends, which is 30 grams per capita per day. These low availabilities are not without nutritional implications, particularly in the feeding of young children. The caloric density of cereal-pulse based diets is low - often around 1 to 1.2 calories per gram of cooked food, and a three-year-old child will have to consume over one kilogram of food to meet its energy needs. Bulk has, in fact, been known to be one of the constraints in satisfying the nutrient requirements of such children, unless the frequency of feeding is increased. A practical and easy way of increasing calorie density is to include fats and sugar - except that these two commodities are today not only in short supply, but also expensive. The recommended per capita intake of each of these food items is 30 grams per day and the effort needed to produce these quantities would be truly enormous.

TABLE. Per Capita Availability of Edible Oils and Sugar in India, 1951-1979

	Edible oils and vanaspati		Sugar	
	kg/year	g/day	kg/year	g/day
1951	2.7	7.7	3.0	8.2
1956	3.2	8.8	5.0	13.7
1961	4.0	10.9	4.7	12.9
1966	3.5	9.6	5.7	15 9
1971	4.5	12.3	7.3	20.0
1976	4.3	11.8	6.2	17.0
1979	4.8	13.1	7.2	19.8

Source: Statistical Outline of India 1980.

TABLE. Per Capita Requirement of Food (g/day) at the National Level on the Basis of Least-cost Balanced Diets

Food	Physiological level	Retail level	Production level
Cereals	386	436	490
Pulses	43	47	53
Leafy vegetables	58	64	72
Other vegetables	45	49	55
Roots, tubers	40	44	50
Milk	200	220	248
Fats, oils	31	34	38
Sugar, jaggery	31	34	38

Source: Recommended Dietary Allowances ICMR 1981.

TABLE. Calculation of National Food Requirements

Per capita production

= Per capita recommended dietary allowance at physiological level

x 1.1 (10 per cent kitchen and other waste)

x 1.125 (for seeds and losses)

Provided that:

Distribution is according to requirements;

There are no constraints on purchasing power.

MEETING NUTRITIONAL NEEDS

At the national level, food production appears to be sufficient to meet the country's needs. In actual practice, however, food consumption does not follow normal distribution but is skewed. A large number of families with a daily income of Rs 2 or less consume diets that do not provide enough energy, and of these, a proportion do not get enough proteins - a finding that explains the widespread PEM among young children. The primary reason for such inequitable distribution is lack of purchasing power. The impressive stocks of food grains, amounting to about 18 million tons, held in recent years, is, in fact, a reflection of this low buying power and consumption. Stocks would have been far less impressive if people could have afforded to buy what they needed. Wages and incomes have gone up over the years but they do not seem to have kept pace with the rising costs of even essential food commodities. Data collected by the National Nutrition Monitoring Bureau show that food consumption has not changed significantly over the last few years.

Due to increased agricultural production in the country, food-grain imports have progressively come down and, during recent years, have all but stopped. The agricultural situation has also been able to prevent the serious widespread famines that used to occur in earlier years. Both are no mean achievements. But increased production seems to have made little impact on the widespread chronic malnutrition in the country, with all its health and developmental implications. The mean birth weight of infants born to mothers belonging to poor rural groups, which was 2.7 kg in 1951, has remained so even today. The proportional mortality rate whereby one-third of all deaths in the country in 1951 occurred among children below the age of five years, continues to be as high even today. Severe forms of protein-energy malnutrition such as kwashiorkor and marasmus were seen in 3 to 5 per cent of children of pre-school age three decades ago. Prevalence rates are not different now. Limited data available on mean body weight and height of adults show little improvement. There are, therefore, no indications that malnutrition has been contained in any way, despite improved food production.

TABLE. Nutritional Status of Pre-school Children

State	Percentage of children				
	Normal	Grade 1	Grade 2	Grade 3	Clinical
Karnataka	9.7	41.6	42.6	6.1	6.6
Maharashtra	7.5	38.2	43.9	10.4	1.5
West Bengal	7.2	40.1	45.1	7.6	4.0
Uttar Pradesh	16.8	40.0	34.2	9.0	1.8

Source: National Nutrition Monitoring Bureau 1977.

Should it be possible to achieve a more equitable consumption of food in the coming years, the level of production, which apparently looks adequate at present, would cease to be so, for two main reasons. Buffer stocks can no longer be maintained at satisfactory levels. Also, children who from birth get enough food, would be able to fully express their genetic potential for growth and become bigger toddlers, bigger adolescents, and bigger adults, with clearly bigger food needs. The level of production would have to increase at a rate faster than that which the country has been able to achieve in the immediate past.

Increased agricultural production is a key factor in ensuring adequate food supplies. The agricultural policy of a country will have to take care of the relevant aspects of its nutrition policy, if the food needs of the population have to be met. Imbalances in production of different commodities have to be corrected and more importantly food has to be made available at a cost that the great majority can afford. Until such time, adequacy of agricultural production will be more apparent than real. It must not be forgotten that factors outside agriculture also have a role in influencing nutrition.

INTERACTION BETWEEN NUTRITION AND HEALTH

Health and nutrition are complementary. While nutrition is recognized as a vital component of health, the latter is an ultimate objective of nutrition. With the widespread prevalence of malnutrition in a given country, no health-care strategy will ever reach the goal of "Health for All" by the year 2000, until and unless nutrition measures are made an important component of health strategy. Similarly, without a sound health-care strategy, nutrition measures, however well conceived and executed, will have very marginal impact. Though there is a growing realization of the importance of health among the nutrition strategists, unfortunately, there is no corresponding appreciation of the importance of nutrition for health strategy. Nutrition is a classical example of a back-bencher among the health sector programmes for several reasons. Fortunately, there is now more dialogue between those responsible for health planning and for nutrition. The efforts to develop nutrition strategy within the framework of primary health care to research the goal of "Health for All" by the year 2000 by the countries of South-East Asia is an excellent example of the appreciation of the close linkage between nutrition and health.

The appreciation of this interaction between nutrition and health has been facilitated by a number of facts. On the one hand, the nutrition planners and programmes became disenchanted with the marginal impact of most nutrition measures, which was explained by several reasons, the two important being the inability of nutrition programmes to reach areas where they are needed most, and the unrealistic inappropriate approach of un-integrated "vertical" programmes. Rational health strategists recognized the health and social implications of malnutrition, and the possibility of having high "health dividends, " if nutrition measures were incorporated. Maternal nutrition and improved birth weight, nutrition and immuno-competence, and several other functional aspects of nutrition are directly related to health programmes.

At the operational level, the question arises as to whether these have produced any significant improvement in the planning and implementation of nutrition and healtn programmes.

THE PAST

It is surprising that, though health and nutrition are complementary and so closely interlinked, the programmes in these two sectors have gone almost on parallel tracks for a very long time. Nutrition has made ceaseless efforts to establish its own identity within the health sector, and health planners did not make much effort to utilize nutrition for health promotion. While human nutrition is recognized as a vital component of health, the latter is the ultimate objective of nutrition. There is general agreement that, without appropriate measures for maintenance and promotion of nutrition, no amount of classical health measures would be effective in ensuring total health. Similarly, in a given country with widespread prevalence of malnutrition, no amount of health planning and health-care strategy will ever reach the ultimate objective. On the other hand, without a sound health-care strategy that includes an appropriate health-care delivery system, nutrition measures, however well-conceived and executed, will have very little chance to go out to those who need them most.

It is interesting to recall that human nutrition promotion through health services, or public health nutrition as it is often designated now, never received priority or was it given even importance among the health sector programmes. In fact, nutrition is a classical example of a back-bencher among various programmes of the health sector for several reasons. Firstly, nutrition means different things to different people. Garry (1953) rightly said that ". . . it is wise to regard it as a meeting-place of the sciences rather than a single scientific discipline." There is real difficulty at present in defining in any precise way what constitutes the field of endeavour in nutrition, let alone priorities within it, for there is a nutrition component in virtually every branch of biology and medicine (Blaxter 1979). It is no wonder that in the past, and even now, nutrition is hardly understood by health administrators, even those at a crucial decision-making level. For a long time, malnutrition was equated with food shortage, and the solution invariably proposed was to produce more food. Thus, when allotting resources in the health sector for various programmes, the tendency was to refer any request for funds for nutrition to other sectors, since control of malnutrition was considered to have had something to do with food and agriculture. The general attitude could be summed up in a sentence: let enough food be produced and malnutrition will automatically disappear.

There is another important reason why nutrition never got recognition as a priority programme of the health sector, even though the dangers of malnutrition were constantly being cited.

While communicable diseases and epidemics immediately drew public attention and forced the hands of the health administrators in allotting priorities and resources, malnutrition with its creeping effects hardly attracted any notice. Moreover, even the large number of deaths in children as a result of malnutrition was always regarded as due to communicable diseases with which they frequently were afflicted with little realization that malnutrition reduces the power of resistance of the victims and leads to infectious disease. malnutrition hardly receives mention as a cause of child mortality in death records.

The Inter-American Mortality Studies in Latin American Countries provides conclusive evidence that in these countries more than 60 per cent of all deaths in children under six years of age is caused by malnutrition directly or indirectly, though the immediate cause of death was infectious disease. Though the study was conducted in the Latin American countries, the result is no doubt equally applicable elsewhere in the developing world. Thus, the creeping nature of malnutrition and its masked effects led to failure to recognize that malnutrition is an important cause of morbidity and mortality. While other health programmes were always accepted as part of the total health strategy, nutrition programmes in the health sector remained foreign. Even when resources for nutrition were to be allocated, for some reason or other no one at the highest level was ever clear as to what was expected of it.

Ironically, the health sector in many developing countries is even now not very clear as to its precise role. In 1977, the World Health Organization (WHO) organized a technical session on nutrition during the World Health Assembly. In the course of the preparation of a background document for this occasion, the WHO sent out a carefully prepared questionnaire to the health ministries of Member States in order to get as much information as possible. When the completed questionnaires were returned, it was clear from the responses that nutrition meant different things to the health sectors of different countries. The approaches adopted differed widely according to what each country thought nutrition should do. This example is being given not as criticism, but because it shows that possibly national nutritionists even now are not very successful in making health administrators aware of the precise role of the health sector in the promotion of nutrition or the prevention of malnutrition.

I would even venture to state that this is apparent not only in the national health administration but in international organizations as well. Take the example of the World Health Organization itself. In 1980/81, it allocated only 1.01 per cent of its regular budget for nutrition despite its official declaration that malnutrition is a priority health problem in the developing world. In contrast, the total budget provision of the Food and Agriculture Organization (FAO) for nutrition in the Food Policy and Nutrition Division came to 2.83 per cent. This is reflected even in the status given to nutrition: WHO headquarters recognizes nutrition as a small unit in the Family Health Division, whereas in the FAO, nutrition along with food policy have the status of a full division.

National health programmes rarely have any nutritional component; nutritional programmes in the health services were developed and implemented as "vertical" programmes, and rarely had any links with other health programmes. During the 1950s many developing countries were formulating various types of nutrition programmes but these were always implemented as isolated activities, with very little dialogue with other programme areas. Instances were not rare when nutrition programmes and maternal and child health programmes in the health sector of a country ran almost on parallel lines, hardly ever converging. A review of past performance reveals the following categories of national nutrition activities (Bagchi 1979):

1. Surveys of dietary intakes and nutritional status.
2. Sponsorship of supplementary feeding programmes with skimmed milk powder (donated by international or bilateral agencies) for women and young children.
3. Distribution of nutrient supplements, e.g. vitamin A and D capsules (donated by international or bilateral agencies) to the vulnerable population.
4. Nutrition education of mothers.

Of these four activities, nutrition surveys can be regarded as the most important. On theoretical considerations, such surveys designed for situation analysis are indeed a basic activity. However, in the majority of cases, such surveys were not followed by action programmes. In other words, the surveys were not planned and executed to obtain data on which programmes could be built, but in fact were continued year after year as an isolated activity. Often the data were not fully analysed and interpreted.

Quite often, the sites for such surveys were selected on the basis of easy accessibility, and not with the objective of obtaining a representative picture of malnutrition. Even in one single country, methods for nutritional assessment varied from one region to another, posing difficulties in a uniform national diagnosis.

Supplementary feeding programmes and distribution of nutrient supplements are two examples of ad hoc activity, designed not for combating a problem, but basically to "do something in nutrition with donated materials." During the 1960s, national health sectors used to receive liberal amounts of milk powder from external agencies, especially UNICEF and CARE, to combat "protein malnutrition," which was then recognized as a problem of great dimensions for which milk was recognized as a definite solution. Distribution of this commodity, of course, was entrusted to hospitals and health centres and, most often, the commodity used to find itself on the "black market" in urban and peri-urban areas.

The population in the peripheral rural areas, who were in greatest need of any form of food supplement, never received any. It is ironic that, even in spite of a radical change in the concept of combating protein-calorie malnutrition, such feeding programmes continued in many countries as an important nutrition activity of the health sector. Distribution of nutrient supplements like vitamin tablets were in fact "misused" because in many cases such tablets were, and are even now, distributed indiscriminately for any form of illness in children and mothers. Frequently, this created an erroneous impression that these nutrients were more important than the consumption of an adequate diet in both quality and quantity. Evaluation of these activities was based on "number of beneficiaries," "weight of milk powder distribution in a month," or "number of vitamin A and D capsules, etc., distributed in a given period. "

Nutrition education is perhaps the most widely used measure in the health sector, with the intention of producing changes in food consumption practices for nutrition improvement. The method, by and large, is by person-to-person or group communication. There are several possible reasons why nutrition education was so commonly adopted by the health services:

1. No capital investment was needed.
2. No specially trained persons were needed. In most cases, this activity was entrusted to nurses, health visitors, or any other health auxiliaries available.
3. It was a suitable way of utilizing the waiting time in out-patient or ante-natal clinics. In most cases, the patients attending such clinics were the "captive audience" for education of this type.
4. There was an erroneous impression that food habits could be changed easily by "lecturing," based on textbook-type messages. The main reason for the lack of evaluation of nutrition education is the "taken-for-granted" attitude about its impact.

In spite of the obvious advantages of the health sector, nutrition education activity was rarely taken seriously as an important task and, in most instances, was included as a routine low-priority activity conducted in a perfunctory manner. While clinical work was always supervised, nutrition education hardly ever was. It is little wonder that nutrition education as a health sector activity had very little impact on the nutritional situation.

There is clearly a need for rationalizing this activity through the health-care system (Bagchi 1977).

THE PRESENT

There is now a rapidly growing realization both among nutritionists and health administrators that there should be a closer dialogue between them. Public health nutritionists have realized that many health programmes have an indirect impact on nutritional status and can be very effective in supplementing their own direct efforts. Even in the absence of specific nutrition interventions, commonly adopted health measures can have appreciable impact on nutritional status (WHO 1980). Improved sanitation and provision of safe drinking-water can significantly reduce gastro-enteritis infectious and parasitic infestations. Immunization can control infectious diseases of childhood. The incidence, duration, and severity of bronchitis and gastro-enteritis could be reduced. Mothers could be encouraged into good child-feeding practices within the limits of their ability to follow them in the face of practical problems of hygiene, food preparation, feeding routines, and availability of the recommended food. Improved child-birth spacing would mean healthier mothers of improved nutritional status who could be better equipped for adequate child care . Programmes for malaria eradication and for expanded immunization are two recent examples of health sector interventions with considerable impact on nutritional status. All these health measures, which are not nutritional activities, can be of tremendous value in moderating the consequences of malnutrition, and even in its prevention.

Nutritionists realized that if direct nutrition measures could be linked with other health programmes of indirect value, then the total effect would be one of considerable benefit to people. This is the basis of the packaged or integrated approach in nutrition. Feeding programmes associated with the provision of safe drinking-water and with education on personal and food hygiene is an example of such an approach.

Wnen the phase of vertical programmes in nutrition failed, it was realized that nutrition is an objective in health; it is not supported by any infrastructure or personnel, nor is this justifiable. For effectiveness, the nutritionists need a delivery system through which nutrition activities can be provided to those who need them most. Since nutrition programmes have no delivery system, it implies that they have to identify programmes or activities that go out to the peripheral areas. Thus nutritionists gradually started integrating nutrition with other health programmes that usually reach peripheral areas, like maternal and child health, family planning, control of communicable diseases, and so on. The historic WHO decision to provide health care to all by the year 2000 through the medium of primary health care has opened up new potentials for nutritionists. Theoretically this is what they have been looking for all these years. Primary health care is an approach with full participation of the community, in which multi-purpose peripheral workers deliver at the same time one, two, or more of the eight elements of the primary health-care system as and when needed. The opportunity is now open for nutrition to be an essential component of primary health care, to be given along with maternal child care, rehydration in diarrhoea, immunization against communicable diseases, advice on birth spacing, etc. Thus, the two important factors limiting the impact of nutrition programmes in the health sector, namely inadequate coverage and an inappropriate approach, can be taken care of through the primary healthcare system. The consultation meeting recently convened by the WHO to develop nutrition strategy within the framework of primary health care in order to reach the goal of health for all by the year 2000 in the countries of the South-East Asia region is an excellent example of appreciation of the interaction between nutrition and health in this part of the world.

Let us examine the extent to which health planners and administrators have appreciated the importance of nutrition in total health development during these years. While health documents and health policies in the past just mentioned malnutrition as a problem almost as a routine, hardly ever in terms of resources and programmes, the present efforts of the health sectors in developing national health strategies to reach the objective of health for all by the year 2000 tell a different story. All countries of the South-East Asia region have included nutrition in the national policy document on health. The national health strategies of most countries have not only mentioned malnutrition as an impediment to total health development, but have identified nutritional indices as parameters for health improvement. In several countries, nutritional indices are being used to assess the quality of life in the total health strategy. In fact, this is possibly the first time that nutrition in its various aspects has been incorporated in national health policies and declarations.

What has led to such increasing recognition by health planners and administrators? The problem of malnutrition and its repercussions on total health is being brought home by nutritionists to national decision-makers. The alarming consequences of four priority nutritional problems, namely proteinenergy malnutrition, nutrition blindness, nutritional anaemia, and endemic goitre, are now the concern not only of nutritionists, but also of national health planners. The health consequences of child malnutrition and their socio-economic implications are of considerable importance to health planners. The health sectors have also taken note of the functional consequences of malnutrition that have been brought into focus in recent years. Its correlation with work-capacity, its association with blindness, and its cause and consequent relationship with infection have all brought home the importance of nutrition to health planners. It is being increasingly realized that nutrition measures if properly implemented might give a high health dividend. Better nutrition means improved maternal nutrition, which leads in turn to improved birth weight, less prematurity, less pregnancy wastage, less infant mortality, and less infant and childhood morbidity. Even in family planning, a programme with the highest priority in many developing countries, it is generally agreed that adoption of family planning would be doubtful without assurance of survival of children, for which adequate nutrition is a vital requirement. The child survival strategy of the family planning programme is indeed based on appropriate promotion of child nutrition.

One must also admit at this stage that the process of health planning has gone through revolutionary changes. The emergence of scientific health planning through prior analysis of the constraints and resources have led to the realization that malnutrition is undoubtedly a constraint and impediment to health development.

THE FUTURE

The future of an integrated nutrition health strategy for the improvement of people's lives is very promising. The strategy of health for all by the year 2000, referred to as a revolutionary idea, is in fact nothing new. However, the unanimous decision of the countries of the world to place AD 200P as a target is undoubtedly a bold step. This decision might not be of much relevance to the developed world where such health care is already within the reach of all, but in a developing country this decision is momentous. Unlike many other decisions in the past, there is already a frantic and concerted effort in most developing countries to give shape to this strategy, and primary health care has already been utilized as one component of the strategy. There is already evidence that countries are taking bold steps in the field of primary health care, and re-ordering their priorities to permeate all levels and sectors concerned with the promotion of health.

It has not been treated as just the addition of yet another layer to the health service (WHO 1981a). Though the terms vary from one country to another and there is no set pattern for primary health care, the common denóminator is the aim of providing essential health care to everyone in the foreseeable future.

It is heartening to note that nutrition occupies an important position in these strategies. The Alma Ata Declaration identified nutrition as one of the eight essential elements of primary health care, but unlike previous declarations this has already come true. Many countries in the developing world have undertaken sizeable projects to determine the means for an effective integration of nutrition with primary health care. A large number of such projects in various centres have provided adequate answers to some vital questions, like the desirable nutritional tasks in the PHC package, identifying simple methods for undertaking these tasks, and the method of training necessary to enable PHC workers to undertake such tasks with competence.

If nutrition is regarded as the meeting-place of various disciplines, then primary health care can be regarded in these new strategies as the meeting-place of nutrition and health. There seems little doubt that future health strategy in developing countries will have nutrition in primary health care as the main weapon with which to fight malnutrition and at the same time provide total health for all.

An important area of future strategy concerns the development of human resources for such a combined nutrition health strategy. The training required for workers at the primary health-care level will be totally different from that provided by conventional methods. In fact, this is already engaging the highest national priority. It is obvious that nutrition trainers must also depart from conventional training, and provide training that would be really and strictly task-oriented. Here again there is frantic activity in many countries and also in international organizations. In WHO's recent document (1981b) outlining the salient points for the training of primary health care workers, there is adequate emphasis on nutrition in the context of total health.

In conclusion one might state that the time has come when nutrition and health are moving towards each other. It is interesting to note that the concepts of health and nutrition have undergone revolutionary changes and, in both, the socio-economic environment is recognized as the most important conditioning factor. The interaction between nutrition and health has possibly reached an ideal equilibrium for the betterment of people's lives.

DISCUSSION

Aspects of Food Production

In Sri Lanka, the range of calorie intake is from 1,859 to 2,600, which is almost exactly the value for the mean plus-minus-two standard deviations . Thus there is a fairly even calorie distribution at all income levels. The policy implication seems to be that if the price of one essential food commodity is subsidized or pegged, good nutrition for the total community will follow.

While good work on the tropical nitrogen cycle has been done in India, this has not been translated into general use. India has set up a tuber crops research institute that deals with production aspects, but in general too little attention had been given to tubers by food scientists, for example, their use in the manufacture of children's foods. Indeed the area of weaning foods is a fertile field for the operation of multinationals in developing countries like Bangladesh.

PROMOTING INTERFACES

How can interface action be promoted in practice? An interregional ministerial group of countries in the region already exists, and has met once in Sri Lanka and again in Nepal. Many conferences, like the one recently held in Manila, have stressed interregional collaboration and exchange. Policy-makers must participate in these meetings to become sensitized to the many aspects of food supply. could not the present group, or an enlarged one, meet a year later, say at Dacca, so as to continue the dialogue? What is needed next is action, now that the need for various interfaces has been so clearly recognized, even when important economic and policy functions are un-represented. So far interactions have occurred almost in spite of individual activities, but now policy planning to positively promote them is needed. Fortunately, food consumption and processing are highly employment-generating activities, and therefore priority areas.

Networks between institutions must be developed. Institutions must be oriented so as to train people to become agents of change. This cannot be left to chance as hitherto but must be purposefully brought about. The United Nations University has started on this road, and can serve as the conscience of international action. The strategies for a hill economy are not identical with those for a river economy. How much and what knowledge is transferable? How can institutions bring about such transfer? How can transferable technologies be generated? Such technologies as rice parboiling or pulse milling have very wide regional application, and must be furnished with institutional backing.

Demand is different from need: how can one increase demand, which is an economic function? Agriculture must be used as a resource to stimulate industrial development. A new science and technology effort must be mounted to build national capabilities in every way, and inter-country cooperation can help to bring this about more rapidly.

Panel Presentations

From Knowledge to Practice in the Home

Practices in the home centre around the members of the family. The mother is still the pace-setter, although the head of the family may appear to be the dominating decision-maker. Practices in the home are influenced by the totality of the interactions between socio-economic, cultural, and educational forces. For the translation of knowledge to practice, knowledge of present practices with regard to food production, food conservation including post-harvest technology, and nutrition, is required. The papers presented so far have thrown light on many of these aspects. If the mother and the members of the family are to adopt the improved practices, we need to strengthen or assure their economic feasibility and viability in the context of the needs of the families, and their capabilities to purchase, adopt, use, and benefit from the new practices. But the major proportion of the families are steeped in poverty, hence the practices have to be built into some economic improvement programmes. The families must be helped to augment their income. The next step is to make available the skills and resources for the changed practices in the most acceptable. appealing, and workable forms. Practices are appealing if a number of leading people follow them. This involves practical training of the home-maker through demonstrations and convincing results.

For any improvement to be adopted on a large scale, women's organizations are essential for leadership, continuity. and feedback. They will then be able to become partners in the development efforts, which are directly related to the improvement of food production, food science, conservation, and nutrition. The infrastructure, and government and voluntary agencies and investments can then be utilized profitably. For that they must have the will and come together. For example, the multimixes and other improved food products can find their way into all the feeding programmes currently in operation. In these programmes, nutrition education must be "built-in." We have numerous evidences for the take-home effects of nutrition education efforts through the pre-school and primary school. All the functions at the village/block level need to get together at their levels and bring about the interaction in that interface. Women must be involved in the decisions at every level.

Formal and non-formal educational channels must be utilized to disseminate the knowledge and bring about the required changes. The undergraduate, post-graduate, and research programmes at the Sri Avinashlingam Home Science College in Coimbatore are practical examples of how this can be done in practice. Thus the B.Sc. Home Science curriculum includes a paper on agriculture, food production, and home science extension education which attempts to enable students to understand the efforts made to augment food production, become aware of the practical work usually done by women in agricultural operations, to know about India's Community Development Programme, National Extension Service and rural agricultural economy, and to understand the role of home science in rural development. The teaching hours are nine per week, and this includes practical sessions and related experience that take the form of actual visits to production centres, co-operatives, fruit farms, dairy farms, poultry farms, local panchayat, mahilar manram, warehouse and land mortgage bank, planning kitchen gardens, starting a bee-keeping unit, preparing aids for teaching rural people, and so on.

The post-graduate programme for the M.Sc. degree includes agriculture, post-harvest technology, and food science taught through four broad areas: food science, nutritional problems in the community, nutrition education, and nutrition intervention programmes. These in turn include such subject areas as agricultural planning, ecological factors, and the roles of food technology, food preservation, and food storage. Placement in a community gives the student an opportunity to observe and learn from direct experience which helps to plan more meaningful education, and maximize nutritional returns by way of food production and storage. A view of the community as a focus of interaction of agriculture, nutrition, food science, and post-harvest technology is thus obtained. Research also seeks to develop such interface areas. Examples are the analysis of new food-grain varieties, analysis of conventional new and uncommon foods, development of low-cost food mixes using local ingredients and of new concepts like leaf protein, community education regarding food storage, and so on.

STORAGE IN PRACTICE

Food-grain management is the overall responsibility of the Food Department of the Government of India. Some facts will indicate the magnitude of the operation. The aim is to maintain a buffer stock of about 12 million tonnes of food grains, besides another 3.5 to 3.8 million tonnes as operational stock for the public distribution system.

Procurement is done by the Food Corporation of India, which also monitors imported grains; the 1979/80 procurement figures were 3.84 million tonnes of rice and 8 million tonnes of wheat.

Public distribution functions through 280,000 ration and fair price shops, of which 77 per cent are in rural areas, and in 1980 some 14.8 million tonnes of total food grains (mainly wheat and rice, but some coarse grains also) were thus handled. Storage of food grains occurs at many levels, and for many purposes. Total storage capacity is about 29 million tonnes (FCI 19.9, Central Warehousing Corporation 3.9, State Warehousing Corporation 5.1). Modern storage structures are designed for either bagged or bulk grains, and are of many types, some with mechanized systems, aeration devices and temperature-measuring facilities. Storage at farmer level is unsatisfactory and both biological and non-biological factors cause tosses of many kinds - physical, financial, hygienic and toxicological.

Loss prevention is an important area of multipronged efforts, which include R&D, training and extension. R&D in storage is pursued by the Indian Grain Storage Institute set up with UNDP financial assistance at Hapur (with five field stations elsewhere) to investigate the nature and extent of losses, improve existing storage structures, develop village-level driers, and train those who will themselves teach or administer food-grain storage. The Central Food Technological Research Institute (CFTRI), Mysore, the Post-harvest Technology Centre, Bhopel, and various agricultural universities also pursue R&D in food-grain storage. As a result of these efforts, losses in handling nearly 20 million tonnes of food grains annually have been less than 1 per cent.

At village level, the Save Grain campaign attempts to educate and motivate farmers through 17 countrywide offices. Demonstration include recommended storage techniques, storage hygiene, periodical inspection, grain cleaning methods and the adoption of model villages. Education and publicity programmes employ numerous media. The scattered offices motivate state governments, voluntary agencies, and village youth organizations, and offer loans and grants towards metal storage bins (of which 250,000 have been sold to farmers), improvement of existing storage structures, and pesticide purchase. Training at all levels is offered at Hapur, by the Food Corporation of India (FCI), the Central Warehousing Corporation, and the CFTRI. Training in food-grain milling is offered at several places.

Processing of paddy is carried out in the country in several types of units, and the out-turn of rice could be improved. Modernization of huller rice mills by adopting rubber roller shells, and setting up efficient mini rice mills have been promoted with success, as have improved parboiling technologies. Improved wheat chakkis are also being promoted.

Legislation to lay down food-grain quality has been enacted to serve as standards for grain for distribution by the public distribution system. The private sector consisting of 200,000 wholesalers and some 3.3 million retailers complements the public distribution system. In all these efforts, it is the collaborative efforts between various government agencies, R&D institutions, voluntary agencies, production units, and so on, that has taken the form of a systems approach to ensuring quality food grains for the people at reasonable prices.

FROM RESEARCH AND DEVELOPMENT TO INDUSTRY

During the past two days the interphases among agriculture, food science, and nutrition have been amplified. The need for constant interaction among breeders, post-harvest technologists, nutritionists, and extension specialists was clearly established. The interaction is very essential to evolve a systems approach leading to increased production levels and optimum utilization without sacrificing nutritional qualities.

Agricultural production as a policy should plan for meeting the requirements of bulk consumption, buffer stock, industry, and export. The main stress, however, will continue to be that of meeting the needs of our large population.

Research and development should focus on the following:

1. Minimizing qualitative and quantitative losses during harvest and post-harvest handling, drying and protection.
2. Devising and utilizing pre-cleaning devices to supply the market with clean and packaged raw produce.
3. Improving milling yields by optimizing appropriate drying and milling procedures, and also by improving milling equipments.
4. Adaptation of processing procedures to ensure freedom from contamination, and good nutrition and acceptability.
5. Evolving techniques of food decontamination and detoxification.
6. By-product utilization.
7. Energy-saving to reduce production cost.
8. Product development and the recommendation of alternate and appropriate technologies designed to meet socio-economic needs. and bring higher returns and purchasing power to the primary producers.
9. Developing and encouraging of primary producers co-operatives to co-ordinate production and processing so that the maximum benefit will accrue to them.

Bibliography

Ausubel, F.M.: *Current Protocols in Molecular Biology*, New York, John Wiley and Sons, 1989.

Brandwein, P.F.: *Sourcebook for the Biological Sciences*, San Diego, Harcourt Brace JOvanovich, 1986.

Brooks, Rodney A.: *Flesh and Machines: How Robots Will Change Us*, New York, Pantheon Books, 2002.

Chrispeels, Maarten : *Plants, Genes and Crop Biotechnology*, Sudbury MA, Jones and Barlett Publishers, 2003.

Cox, Earl D. : *Beyond Humanity: Cyber Evolution and Future Minds*, Roackland, Charles River Media, 1996.

David Sadava: *Plants, Genes and Crop Biotechnology*, Sudbury MA, Jones and Barlett Publishers, 2003.

Dixon, Dougal: *After Man-A Zoology of the Future*, New York, St. Martin's Press, 1981.

Dodds, John H.: *Plant Genetic Engineering*, New York, Cambridge University Press, 1985.

Drexler, K. Eric: *Engines of Creation: The Coming Era of Nanotechnology*, New York, Anchor Books, 1986.

Fransman M, Junne G, Roobeek A: *The Biotechnology Revolution?*, Oxford, Blackwell, 1995.

Friedberg, E.C.: *DNA Repair*, New York, WH Freeman and Company, 1985.

Fumento, Michael: *Bioevolution: How Biotechnology is Changing Our World*, San Francisco, Encounter Books, 2003.

Gaskell G: *Biotechnology-the Making of a Global Controversy*, Cambridge, Cambridge University Press, 2002.

George S. Paul: *Beyond Humanity: Cyber Evolution and Future Minds*, Roackland, Charles River Media, 1996.

Goodman, D.C.: *From Farming to Biotechnology: A Theory of Agro-industrial Development*, Oxford, Blackwell, 1987.

Goodsell, David S.: *Bionanotechnology: Lessons From Nature*, Hoboken, Wiley-Liss, 2004.

Kurzweil, Ray: *The Age of Spiritual Machines*, New York, Penguin Books, 1999.

Moravec, Hans: *Mind Children: The Future of Robot and Human Intelligence*, Cambridge, Harvard University Press, 1988.

Morholt, E. : *Sourcebook for the Biological Sciences*, San Diego, Harcourt Brace JOvanovich, 1986.

Murray, David: *Seeds of Concern: The Genetic Manipulation of Plants*, Sydney, University of New South Wales, 2003.

Old, R.W. : *Principles of Gene Manipulation*, London, Blackwell Scientific Publications, 1989.

Postman, Neil: *Technopoly: The Surrender of Culture to Technology*, New York, Vintage Books, 1992.

Primrose, S.B.: *Principles of Gene Manipulation*, London, Blackwell Scientific Publications, 1989.

Retzer, W.J.: *Biotechnology Workbook*, Englewood Cliffs, Prentice Hall, 1991.

Rifkin, Jeremy: *The Biotech Century*, New York, Penguin Putnam, 1998

Shetty, Kalidas: *Food Biotechnology*, New York, Dekker/CRC Press, 2005.

Slater, R.J.: *Experiments in Molecular Biology*, Clifton, Humana Press, 1986.

Smith, John E.: *Biotechnology*, Cambridge, Cambridge University Press, 2004.

Switzer, R.L.: *Experimental Biochemistry*, New York, W.H. Freeman and Company, 1977.

Towle, Albert: *Modern Biology*, Austin, Holt, Rinehart and Winston, 1988.

Walden, Richard: *Genetic Transformation in Plants*, England, Open University Press, 1988.

Winston, Mark L.: *Travels in the Genetically Modified Zone*, Cambridge, Harvard University, 2004.

Index

□□□